COMBINATORICS

Set Systems, Hypergraphs, Families of Vectors and Combinatorial Probability

BÉLA BOLLOBÁS
University of Cambridge and Louisiana State University

PUBLISHED BY THE PRESS SYNDICATE OF THE UNIVERSITY OF CAMBRIDGE
The Pitt Building, Trumpington Street, Cambridge CB2 1RP, United Kingdom

CAMBRIDGE UNIVERSITY PRESS
The Edinburgh Building, Cambridge CB2 2RU, UK http://www.cup.cam.ac.uk
40 West 20th Street, New York, NY 10011–4211, USA http://www.cup.org
10 Stamford Road, Oakleigh, Melbourne 3166, Australia

First published 1986
Reprinted 1988, 1990, 1994, 1998

A catalogue record for this book is available from the British Library

Library of Congress Cataloguing in Publication Data

Bollobás, Béla
Combinatorics: set systems, hypergraphs, families of vectors and combinatorial probability
Bibliography: p.
Includes index.
1. Combinatorial analysis. I. Title.
QA164.B65 1986 511′.6 86-9602

ISBN 0 521 33703 8 paperback

Transferred to digital printing 2001

COMBINATORICS

From a print in Sir Isaac Newton's copy of *Traité de Combinaisons*, by Remond de Monmort, Paris, 1713.

To the memory of my Father

'A MATHEMATICIAN, like a painter or a poet, is a maker of patterns. If his patterns are more permanent than theirs, it is because they are made with *ideas*.' 'The mathematician's patterns, like the painters's or the poet's, must be *beautiful*; the ideas, like the colours or the words, must fit together in a harmonious way. Beauty is the first test: there is no permanent place in the world for ugly mathematics.'

G. H. HARDY

A Mathematician's Apology

CONTENTS

PREFACE

This book is an expanded account of a first-year graduate course in combinatorics, given at Louisiana State University, Baton Rouge during the fall semester of 1985.

The traditional ingredients of an initial combinatorics course seem to be combinatorial identities and generating functions, with some introductory design theory or graph theory. Usually, set systems, hypergraphs and sets of vectors are barely mentioned. But these topics are as worthy of consideration as any, in view of their fundamental nature and elementary structure.

The properties of the subsets of a finite set must lie at the heart of any study of combinatorial theory. This was the motivation for the course given at LSU. It is hoped that combinatorics courses of this nature will be given from time to time, alternating with others of a more standard kind. The book contains considerably more material than one could reasonably hope to cover in a one semester course: this gives the lecturer ample freedom to slant the lectures to his taste.

The book is aimed at future combinatorialists and other mathematicians alike — in addition to the combinatorialists, those specializing in analysis and probability theory are most likely to find the material stimulating.

The contents of the book are made fairly clear by the list of section headings. We look at the very basic features of subsets, such as containment, disjointness, or specified intersection and partition properties. Many of the results have immediate and beautiful applications in combinatorial probability and in analysis. In addition to a thorough grounding in the theory of set systems, the reader will be introduced to some of the basic results in matroid theory, the theory of designs and the Ramsey theory of infinite sets. The reader can consolidate his understanding of the material by tackling over one hundred classroom-tested exercises, ranging from the simple to the challenging. This treatment is by no means comprehensive; it is inevitable that the selected material reflects the taste of the author. Emphasis has been given to theorems with elegant and beautiful proofs; those which may be called the gems of the theory.

Many of the theorems have considerable extensions often of a rather technical nature. In order to preserve the leisurely pace and chatty style, these extensions are not presented. Accordingly, the book's hundred or so references are also far from being comprehensive.

Very few special terms are employed in this subject and all the necessary ones are defined. However, some concepts from graph theory are mentioned in passing, occasionally without explanation, to avoid introducing unnecessarily dull passages into the narrative. These concepts are for illustration, though, and are not essential. In any case it is expected that most readers of the book and most students attending a course on it will have encountered elementary graph theory beforehand.

It is a great pleasure to thank the participants of the course at Baton Rouge, including my seven Ph.D. students at Cambridge, for their stimulating comments and the lively atmosphere during the lectures. I have especially benefited from the help of Keith Ball, Graham Brightwell, Hugh Hind and Imre Leader.

The book could not have been produced in such a short time without the generous help from Dr. Andrew Harris, to whom I am very grateful.

Finally, I would like to thank my son, Mark, for helping whenever possible, and my wife, Gabriella, for her continuous encouragement, support and love.

B.B.

§1. NOTATION

All we shall do in this brief section is define the basic concepts and introduce some of the notation we shall need in the book. It is rather fortunate that remarkably few definitions are needed to start work in the subject; further definitions will be given when needed. The number of elements in a set Y is denoted by $|Y|$. The *power set* of Y is the set of all subsets of Y: it is denoted by $\mathcal{P}(Y)$. The set of all r-subsets, i.e. subsets of size r, is denoted by $Y^{(r)}$. In §20 we shall consider infinite sets but in the rest of the book we shall always work with finite sets. In fact, X will always stand for a set of size $n \geq 1$, and we shall often identify X with $[n] = \{1, 2, \ldots, n\}$.

Partially ordered sets will not be studied explicitly in this book; however, as they are relevant to the material in several sections, we shall define them here. A *partially ordered set* or briefly a *poset* is a pair $(S, \leq)$ such that S is a set and $\leq$ is a transitive binary relation on S, i.e. $x \leq y$ and $y \leq z$ imply $x \leq z$, such that $x \leq x$, and $x \leq y$ and $y \leq x$ imply $x = y$. Defining "$x < y$" to mean that "$x \leq y$ and $x \neq y$", we find that $<$ is a transitive and irreflexive binary relation on S. Conversely, if $<$ is a transitive and irreflexive binary relation on S then letting "$x \leq y$" mean "$x < y$ or $x = y$" we get a poset $(S, \leq)$. In view of this, we may talk of a poset $(S, <)$. Not surprisingly, "$x > y$" means "$y < x$", and "$x \geq y$" is the same as "$y \leq x$".

With the natural order, namely $A \leq B$ if $A \subset B$, the set $\mathcal{P}(Y)$ is a poset. In fact, $\mathcal{P}(Y)$ is a *lattice*: for any two sets A and B, there is a smallest set which is at least as large as either, namely $A \cup B$, and there is a largest set which is at least as small as either, namely $A \cap B$. The partially ordered set $\mathcal{P}(Y)$ is the disjoint union of its level sets $Y^{(r)}$.

A *set system* (*on* Y) is a pair $(Y, \mathcal{F})$, where Y is a set and $\mathcal{F} \subset \mathcal{P}(Y)$, and an *$r$-graph* or *$r$-uniform hypergraph* (*on* Y) is a pair $(Y, \mathcal{E})$ where $\mathcal{E} \subset Y^{(r)}$. The set Y is the *ground set*. An element of $\mathcal{E}$ is a *hyperedge* or simply an *edge* of the hypergraph. A *graph* is a 2-graph: to distinguish it from a hypergraph it may be called a *linear graph* or an *edge graph.* In a graph a point x is said to be *joined* to y if $\{x,y\}$ is an edge of the graph. A point of a graph is usually called a *vertex* and an edge $\{x,y\}$ is written as xy or yx. We say that the vertex x is *incident* with the edge xy and the vertices x and y are *adjacent.* Occasionally the points of the ground set of a hypergraph are also called vertices. The ground set of a set system or hypergraph is usually taken to be $X = [n]$ and is often omitted from the notation; thus we shall talk of a set system $\mathcal{F}$ and a hypergraph $\mathcal{E}$. The *size* of a set system $(X, \mathcal{F})$ is $|\mathcal{F}|$, the number of sets is $\mathcal{F}$ and the *order* is $|X|$, which is usually n. The same terminology is used for hypergraphs and graphs. Instead of a set system $\mathcal{F}$ on X we shall sometimes talk of a *family $\mathcal{F}$ of subsets of* X or simply a *family* $\mathcal{F}$. We shall call a family $\mathcal{F}$ *isomorphic* to a family $\mathcal{G}$ if there is a 1-1 map from $\bigcup_{F\in\mathcal{F}} F$ to $\bigcup_{G\in\mathcal{G}} G$ setting up a 1-1 correspondence between the sets in $\mathcal{F}$ and the sets in $\mathcal{G}$. Thus if $\mathcal{F} = \{12, 123, 124, 2345\}$ and $\mathcal{G} = \{267, 27, 237, 2346\}$ then $\mathcal{F} \cong \mathcal{G}$. Here, for the sake of simplicity, we have written $ab\dots u$ for the set $\{a, b, \dots, u\}$; this will be done in future when there is no danger of confusion. A subset A of X is naturally identified with the characteristic function χ_A of A on X:

$$\chi_A(x) = \begin{cases} 1 & \text{if } x \in A \\ 0 & \text{if } x \in X \setminus A. \end{cases}$$

Then $\mathcal{P}(X)$ is identified with $\{0,1\}^X$, the set of all functions from X to $\{0,1\}$. Also, under the identification $X = [n]$, a 0-1 function on X is a 0-1 sequence of length n. Occasionally it is convenient to consider $\{0,1\}$ endowed with addition and multiplication, turning it into Z_2, the field of order 2. Under this correspondence the intersection $A \cap B$ of two sets A and B corresponds to the product of the corresponding functions. What about the sum of the functions? It corresponds to the *symmetric difference* $A \triangle B$ of the sets: $A \triangle B = (A \setminus B) \cup (B \setminus A)$, where $A \setminus B$ is the usual difference of sets: $A \setminus B = \{a \in A : a \notin B\}$. Incidentally, this correspondence shows that the symmetric difference operation is associative and commutative: $(A \triangle B) \triangle C = A \triangle (B \triangle C)$ and $A \triangle B = B \triangle A$. Furthermore, $A \triangle \emptyset = A$, $A \triangle A = \emptyset$ and $(A \triangle B) \triangle (B \triangle C) = A \triangle (B \triangle B) \triangle C = A \triangle C$.

We often associate a graph with a poset. An element x of a poset $P = (S, <)$ is said to *cover* an element y if $x > y$ and no element z satisfies $x > z > y$. The *covering graph* of P has vertex set S and x is joined to y if x covers y or y covers x.

In particular, the covering graph of $\mathcal{P}(X)$ (with its natural order) has vertex set $\mathcal{P}(X)$ and a set A is joined to a set B if $A \triangle B$ has precisely one element, i.e. either $A = B \cup \{x\}$ or $B = A \cup \{x\}$ for some $x \in X$ (and $A \neq B$). This graph is precisely the graph of the *n-dimensional cube*, which we shall denote by Q^n. It is more customary to construct Q^n on the set of all 2^n 0-1 sequences of length n by joining two sequences if they differ in exactly one place.

When studying $\mathcal{P}(X)$, it is often expedient to look at Q_n because this graph, *without* the labels of the vertices, reflects all the essential properties of $\mathcal{P}(X)$. In fact, if we pick any of the vertices of Q^n and let it correspond to the empty set $\emptyset$ or the entire ground set X then the labels are uniquely determined, up to a permutation on X. Thus the study of set systems is just the study of subsets of vertices of Q^n. Of course, this almost empty statement is not supposed to help in proving results about set systems.

The *Hamming distance* or simply the *distance* between two sets A and B is $d(A, B) = |A \triangle B|$. This is just the distance in the graph Q^n between the corresponding vertices; it is also the number of places in which the corresponding 0-1 sequences differ.

As customary, we shall write $\lfloor x \rfloor$ for the *floor* of a real number x, namely for the greatest integer not greater than x, and we write $\lceil x \rceil$ for the *ceiling* of x, i.e. for $-\lfloor -x \rfloor$, the smallest integer not smaller than x. Finally, $(x)_r$ denotes the *falling factorial*: $(x)_r = x(x-1)\dots(x-r+1)$.

§2. REPRESENTING SETS

We shall start with a playful and simple problem. Given a set system $(X, \mathcal{F})$, let us try to replace X by a smaller set Y which still distinguishes the elements of $\mathcal{F}$. In other words, let us try to find a set $Y \subset X$, $Y \neq X$, such that $A \cap Y \neq B \cap Y$ for $A, B \in \mathcal{F}$, $A \neq B$. Of course, it suffices to look for a set Y of the form $Y = X - \{x\}$, i.e. it suffices to find an element $x \in X$ such that $A - \{x\} \neq B - \{x\}$ whenever $A, B \in \mathcal{F}$, $A \neq B$. If $\mathcal{F}$ has small size then this can certainly be done, but if $\mathcal{F}$ is large such a Y need not exist. The following result, due to Bondy (1972), shows that the critical value of $|\mathcal{F}|$ is precisely n.

Theorem 1. *Let $\mathcal{F} = \{A_1, A_2, \ldots, A_n\}$ be a set system on $X = [n]$. Then there is an element $x \in X$ such that $A_1 - \{x\}$, $A_2 - \{x\}$, $\ldots$, $A_n - \{x\}$ are all distinct. The set system $\mathcal{F} = \{\emptyset, \{1\}, \{2\}, \ldots, \{n\}\}$ shows that such an x need not exist if $|\mathcal{F}| = n + 1$.*

Proof. Suppose the set system $\mathcal{F}$ is a counterexample to the assertion. Then $n \geq 2$ and our aim is to arrive at a contradiction. We shall give two different ways of doing this.

For the first proof, set $\mathcal{D} = \{D \subset X : |\mathcal{F}_D| \geq |D| + 1\}$, where $\mathcal{F}_D = \{D \cap A_i : i \in \mathcal{F}\}$. If $A_1, A_2 \in \mathcal{F}$ and $d \in A_1 \triangle A_2$ then $\{d\} \in \mathcal{D}$ so $\mathcal{D} \neq \emptyset$. Let D be a maximal set in $\mathcal{D}$. Then $|D| \leq n - 2$ and $|\mathcal{F}_D| \leq n - 1$ (so, in fact, $|\mathcal{F}_D| = n - 1$).

As $|\mathcal{F}_D| \leq n-1$, there are sets $A_1, A_2 \in \mathcal{F}$ such that $A_1 \cap D = A_2 \cap D$. Let $e \in A_1 \triangle A_2$ and set $E = D \cup \{e\}$. Since $A_i \cap D \neq A_j \cap D$ implies that $A_i \cap E \neq A_j \cap E$, the system $\mathcal{F}_E$ has at least one more set than $\mathcal{F}_D$:

$$|\mathcal{F}_E| \geq |\mathcal{F}_D| + 1 \geq |D| + 2 = |E| + 1.$$

Therefore $E \in \mathcal{D}$, contradicting the maximality of D.

For the second proof note first that if $A, B \subset X$, $A \neq B$ and $A - \{i\} = B - \{i\}$ then either $A = B \cup \{i\}$ or $B = A \cup \{i\}$. Hence, in either case, $A \triangle B = \{i\}$.

The set system $\mathcal{F}$ is such that for every $i \in X$ there are $k = k(i)$ and $l = l(i)$, $1 \leq k < l \leq n$, such that $A_k - \{i\} = A_l - \{i\}$, i.e. $A_k \triangle A_l = \{i\}$. Construct a graph H on $[n] = \{1, 2, \ldots, n\}$ by joining $k(i)$ to $l(i)$, $i = 1, 2, \ldots, n$. We claim that H does not contain a cycle, i.e. a sequence $a_1, a_2, \ldots, a_s$ of distinct vertices such that a_j is joined to a_{j+1}, where, by definition, $a_{s+1} = a_1$. Indeed, otherwise we may assume that $a_j = j$, $k(j) = a_j$ and $l(j) = a_{j+1}$, $j = 1, 2, \ldots, s$. But then

$$\{s\} = A_1 \triangle A_s = (A_1 \triangle A_2) \triangle (A_2 \triangle A_3) \triangle \ldots \triangle (A_{s-1} \triangle A_s)$$

$$\subset \bigcup_{j=1}^{s-1} (A_j \triangle A_{j+1}) = \{1, 2, \ldots, s-1\}$$

a blatant contradiction!

Therefore, H has n vertices, n edges and no cycles. However, this is impossible since a simple result in graph theory states that if a graph has n vertices, m edges and no cycles then $n \leq m - 1$ (see Ex. 1). This completes the second proof. ■

In §17 we shall give yet another proof of this result: we shall deduce it from a more general theorem (see Exercise 17.1).

The next result we consider, Hall's marriage theorem, is one of the basic results in combinatorics and has many far-reaching consequences. It resembles the light-hearted problem we have just discussed, but this resemblance is purely superficial. Given a set system $\mathcal{F} = \{A_1, A_2, \ldots, A_m\}$ on X, when can we find m distinct elements of X, one from each A_i, $i = 1, 2, \ldots, m$? Such a set of m distinct elements is called a *set of distinct representatives* or a *transversal* of the system $\mathcal{F}$. Note that if $\{x_1, x_2, \ldots, x_m\}$ is a set of distinct representatives and $S \subset [m] = \{1, 2, \ldots, m\}$ then

$$\left| \bigcup_{i \in S} A_i \right| \geq \left| \bigcup_{i \in S} \{x_i\} \right| = \left| \{x_i : i \in S\} \right| = |S|.$$

The beautiful result that this trivial necessary condition is, in fact, sufficient to ensure the existence of a set of distinct representatives, is usually referred to as *Hall's theorem* (1935) or *Hall's marriage theorem*, though an equivalent form of it had been discovered earlier by König (1931) and the result is also a special case of Menger's theorem (1927). For many related results see Bollobás (1978, Ch. II) and (1979, Ch. III).

Theorem 2. *A set system $\mathcal{F} = \{A_1, A_2, \ldots, A_m\}$ has a set of distinct representatives iff*

$$\left|\bigcup_{i \in S} A_i\right| \geq |S| \tag{1}$$

for all $S \subset [m]$.

Proof. We have already noted the necessity of (1), called *Hall's condition*, so let us turn to the proof of the sufficiency. There are many elegant proofs; here we choose one using induction on m. For $m = 1$ the assertion is trivial so suppose that $m > 1$ and the result holds for smaller values of m. Let $\mathcal{F}$ be a system satisfying (1).

Suppose first that for $S \subset [m]$, $S \neq \emptyset$ and $S \neq [m]$, we have strict inequality in (1). Pick an element $x_m \in A_m$ and set $A'_i = A_i \setminus \{x_m\}$, $i = 1, \ldots, m-1$. Then the system $\{A'_1, A'_2, \ldots, A'_{m-1}\}$ also satisfies Hall's condition so it has a set of distinct representatives, say $\{x_1, x_2, \ldots, x_{m-1}\}$. Then $\{x_1, x_2, \ldots, x_m\}$ is a set of distinct representatives of $\mathcal{F}$.

Suppose now that there is a set $S \subset [m]$, $S \neq \emptyset$ and $S \neq [m]$, for which equality holds in (1). Set $\mathcal{F}_1 = \{A_i : i \in S\}$. By our induction hypothesis the system $\mathcal{F}_1$ has a set of distinct representatives, say X_1. Clearly $X_1 = \bigcup_{i \in S} A_i$, $|X_1| = |S|$. For $i \in [m] \setminus S$ set $A'_i = A_i \setminus X_1$ and let $\mathcal{F}_2 = \{A'_i : i \in [m] \setminus S\}$. Does $\mathcal{F}_2$ satisfy Hall's condition? Certainly, since if $T \subset [m] \setminus S$, $T \neq \emptyset$, then

$$\left|\bigcup_{i \in T} A'_i\right| \geq \left|\bigcup_{i \in T} A_i \cup X_1\right| - |X_1| = \left|\bigcup_{i \in S \cup T} A_i\right| - |S| \geq |T|.$$

Hence, by the induction hypothesis, $\mathcal{F}_2$ also has a set of distinct representatives, say X_2. Then $X_1 \cup X_2$ will do for $\mathcal{F}$.

■

Theorem 2 is called the marriage theorem because an easy reformulation of it claims that we can marry each of m girls to a boy she knows iff any k girls know at least k boys. This can be stated more precisely in terms of matchings in bipartite graphs. A ***bipartite graph with bipartition*** (V_1, V_2) is a graph with vertex set $V_1 \cup V_2$ such that $V_1 \cap V_2 = \emptyset$ and every edge joins a vertex of V_1 to a vertex of V_2. A *matching* from V_1

to V_2 is a set of edges such that every vertex of V_1 is incident with precisely one edge and every vertex of V_2 is incident with at most one edge. Write $\Gamma(S)$ for the set of vertices adjacent to vertices in S. Note that there is a matching from V_1 to V_2 iff the sets $\Gamma(x)$, $x \in V_1$, have distinct representatives. Thus, Theorem 2 has the following reformulation.

Theorem 2′. *In a bipartite graph with bipartition (V_1, V_2) there is a matching from V_1 to V_2 iff*

$$|\Gamma(S)| \geq |S| \tag{1'}$$

for all sets $S \subset V_1$. ∎

If Hall's condition fails then we cannot find a set of distinct representatives. But how close can we come to finding one? When can we find distinct representatives for all but d of them? Clearly only if the union of any k sets has at least $k - d$ elements. Once again, this trivial necessary condition is, in fact, sufficient. This is the defect form of Hall's theorem.

Corollary 3. *Suppose the sets $A_1, A_2, \ldots, A_m$ are such that*

$$\left|\bigcup_{i \in S} A_i\right| \geq |S| - d$$

for all $S \subset [m]$. Then we can find distinct representatives for all but d of the sets A_i.

Proof. Let D be a set of size d, disjoint from $\bigcup_{i=1}^{m} A_i$, and set $A_i' = A_i \cup D$, $i \in [m]$. Then the system $\{A_1', A_2', \ldots, A_m'\}$ satisfies Hall's condition so it has a set of distinct representatives. All but at most d elements of this set belong to the original A_i's. ∎

Let us conclude this section with a consequence of Hall's theorem concerning the level sets $X^{(r)}$ of the partially ordered set $\mathcal{P}(X)$.

Corollary 4. *For $r < n/2$ there is an injection (a 1-1 function) $f_r : X^{(r)} \to X^{(r+1)}$ such that $A \subset f(A)$. For $r > n/2$ there is an injection $g_r : X^{(r)} \to X^{(r-1)}$ such that $A \supset g(A)$.*

Proof. Replacing each set by its complement in X, the second assertion follows from the first so it suffices to prove the first. Let then $r < n/2$ and consider the bipartite graph with classes $V_1 = X^{(r)}$ and

$V_2 = X^{(r+1)}$ in which $A \in X^{(r)}$ is joined to $B \in X^{(r+1)}$ iff $A \subset B$. All the corollary claims is that there is a matching from V_1 to V_2. Is then condition (1′) satisfied? Certainly, since for $S \subset V_1$ every $A \in S$ is joined to $n - r$ sets $B \in \Gamma(S)$ and every $B \in \Gamma(S)$ is joined to at most $r + 1$ sets $A \in S$, so

$$|S|(n-r) \leq \left|\{AB : A \in S, B \in V_2, A \subset B\}\right| \leq |\Gamma(S)|(r+1). \qquad (2)$$

Hence, by Theorem 2′, there is indeed a matching from V_1 to V_2.

Exercises

1. Prove by induction on the number of vertices that if an edge graph has no cycles then it has fewer edges than vertices.

2. Prove that if an edge graph has n vertices, $n - 1$ edges and no cycles then it is connected. (Such a graph is called a *tree*.)

3. Consider a tree with vertex set $[n+1]$ and n edges. Orient the edges and label them with $1, 2, \ldots, n$, with any two edges getting distinct labels (see Figure 1). Show that there is a unique set system $\{A_1, A_2, \ldots, A_{n+1}\}$ on $[n]$ such that if kl is an edge oriented from k to l and labelled i then $A_l = A_k \cup \{i\}$.

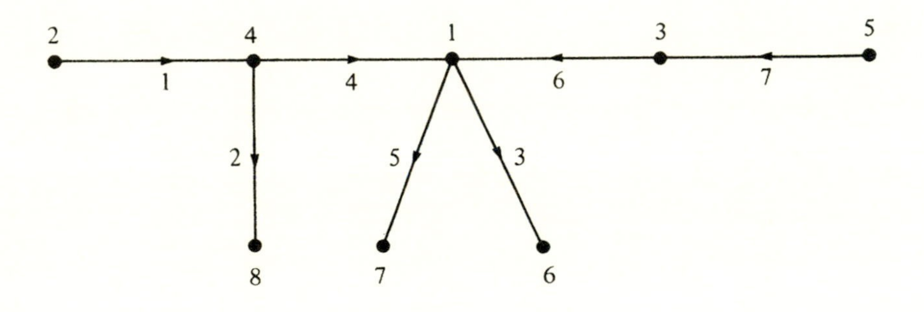

Figure 1. *An oriented and labelled tree of order* 8. *This tree gives the set system* $A_1 = \{1,4,6,7\}$, $A_2 = \{6,7\}$, $A_3 = \{1,4,7\}$, $A_4 = \{1,6,7\}$, $A_5 = \{1,4\}$, $A_6 = \{1,3,4,6,7\}$, $A_7 = \{1,4,5,6,7\}$, $A_8 = \{1,2,6,7\}$.

4. Show that the following variant of Theorem 1 holds for the edge set $\mathcal{F}$ of a graph.

Let $\mathcal{F} \subset X^{(2)}$ be such that if $Y \subset X$, $|Y| = n-2$, then there exist distinct sets $F_1, F_2 \in \mathcal{F}$ inducing the same subset of Y: $F_1 \cap Y = F_2 \cap Y$. Then $|\mathcal{F}| \geq \lceil 3(n-1)/2 \rceil$.

Show also that for each $n \geq 3$ there is a system $\mathcal{F}$ of size $\lceil 3(n-1)/2 \rceil$ satisfying the conditions.

5. Let A be an n by n 0-1 matrix. Prove that A has n 1's such that each row and each column contains precisely one of them iff any k rows contain 1's in at least k columns. (König (1931))

6. Let G be a bipartite graph with bipartition (V_1, V_2). Suppose every vertex in V_1 has degree at least k (i.e. has at least k vertices adjacent to it) and every vertex in V_2 has degree at most k. Check that the proof of Corollary 4 shows that in G there is a matching from V_1 to V_2.

7. Prove the following reformulation of Corollary 4: for $r \leq \lceil n/2 \rceil$ there is a surjection $f_r : X^{(r)} \to X^{(r-1)}$ such that $A \supset f_r(A)$ and for $r \geq \lfloor n/2 \rfloor$ there is a surjection $g_r : X^{(r)} \to X^{(r+1)}$ such that $A \subset g_r(A)$.

§3. SPERNER SYSTEMS

A set system $\mathcal{F}$ is said to be a *Sperner system* or a *Sperner family* if no set in it contains another: $A, B \in \mathcal{F}$ and $A \neq B$ imply $A \not\subset B$. Thus, $\mathcal{F}$ is a Sperner system if it consists of incomparable elements of the partially ordered set $\mathcal{P}(X)$.

Perhaps the simplest examples of Sperner families are the level sets of $\mathcal{P}(X)$, namely $X^{(0)}, X^{(1)}, \ldots, X^{(n)}$, a largest of which is $X^{(\lfloor n/2 \rfloor)}$. The reason for the terminology is the basic and rather simple result of Sperner (1928) stating that no Sperner family can have more elements than $X^{(\lfloor n/2 \rfloor)}$.

Theorem 1. *A Sperner system $\mathcal{F}$ on $X = [n]$ consists of at most $m = \binom{n}{\lfloor n/2 \rfloor}$ sets.*

Proof. We claim that $\mathcal{P}(X)$ can be partitioned into m chains, i.e. systems of the form $\{A_{i_1}, A_{i_2}, \ldots, A_{i_t}\}$ where $A_{i_1} \subset A_{i_2} \subset \ldots \subset A_{i_l}$. Indeed, such chains can be built from the functions f_r, g_r in Corollary 2.4: for $A \in X^{(r)}$ and $r < \lfloor n/2 \rfloor$ put A and $f_r(A)$ into the same chain and for $B \in X^{(r)}$ and $r > \lfloor n/2 \rfloor$, put B and $g_r(B)$ into the same chain. The pairs $A \subset f_r(A)$ and $B \supset g_r(B)$ induce a partition of $\mathcal{P}(X)$ into disjoint chains such that each chain contains a member of $X^{(\lfloor n/2 \rfloor)}$, of which we have exactly m. The assertion of the theorem is immediate since $\mathcal{F}$ meets every chain $A_{i_1} \subset A_{i_2} \subset \ldots \subset A_{i_l}$ in at most one element. ∎

The proof above is by no means the simplest and neither is it the most elegant, but it is, perhaps, the most natural one (see also Ex. 1). A considerably sharper result, Theorem 2 below, is due to Lubell (1966). The same result was discovered earlier by Meshalkin (1963) and, with some effort, it can also be read out of some results of Yamamoto (1954). Though Lubell's result is also a rather special case of an earlier lemma of Bollobás (1965), inequality (1) below has become known as the *LYM inequality*.

Theorem 2. *Let $\mathcal{F}$ be a Sperner system on $X = [n]$ and set $\mathcal{F}_k = \mathcal{F} \cap X^{(k)}$ and $f_k = |\mathcal{F}_k|$. Then*

$$\sum_{k=0}^{n} f_k \binom{n}{k}^{-1} \leq 1. \tag{1}$$

Proof. A maximal chain in $\mathcal{P}(X)$ is of the form $\emptyset \subset A_{i_1} \subset A_{i_2} \subset \ldots \subset A_{i_{n-1}} \subset X$, where $A_{i_k} \in X^{(k)}$. There are $n!$ maximal chains since every $A \in X^{(k)}$ contains precisely k members of $X^{(k-1)}$. Every chain contains at most one element of $\mathcal{F}$ and every set $A \in \mathcal{F}_k$ is contained in $k!(n-k)!$ maximal chains. Hence

$$\sum_{k=0}^{n} f_k k!(n-k)! \leq n!. \qquad \blacksquare$$

Occasionally, Theorem 2 is formulated in terms of weights. Let the *weight* of a set $A \in X^{(a)}$ be $w(A) = \binom{n}{a}^{-1}$.

Theorem 2′. *Let $\mathcal{F} = \{A_1, A_2, \ldots, A_m\}$ be a Sperner system on X. Then*

$$\sum_{i=1}^{m} w(A_i) \leq 1. \tag{1'}$$

$\blacksquare$

Lubell's proof of Theorem 2, reproduced above, is very elegant indeed. It has also an equally elegant reformulation. For a set $A \in X^{(a)}$ let us say that a permutation $x_1 x_2 \ldots x_n$ of X *contains* A if $\{x_1, x_2, \ldots, x_a\} = A$. Note that A is contained in precisely $a!(n-a)! = w(A)n!$ permutations. Now if $\{A_1, A_2, \ldots, A_m\}$ is a Sperner system then each of the $n!$ permutations contains at most one A_i. Consequently

$$\sum_{i=1}^{m} w(A_i) n! \leq n!,$$

implying (1′). To recover the original proof, simply identify a permutation $x_1 x_2 \ldots x_n$ with the maximal chain $\emptyset \subset \{x_1\} \subset \ldots \subset \{x_1, x_2, \ldots, x_n\} = X$.

What the LYM inequality tells us is that if we want to construct a large Sperner system then we'd better choose sets of size about $n/2$. We cannot do better than choose sets whose weights are as small as possible i.e. sets of sizes $\lfloor n/2 \rfloor$ or $\lceil n/2 \rceil$. This latter assertion is precisely Theorem 1.

For what families do we have equality in (1)? We certainly have equality for $X^{(r)}$, $r = 0, 1, \ldots, n$. The following result, which can be considered to be a local LYM inequality, shows that these are the only ones.

Theorem 3. *Let $0 < r \le n$, $\mathcal{A} \subset X^{(r)}$ and set*

$$\partial\mathcal{A} = \{B \in X^{(r-1)} : B \subset A \textit{ for some } A \in \mathcal{A}\}.$$

Then

$$|\partial\mathcal{A}| \Big/ \binom{n}{r-1} \ge |\mathcal{A}| \Big/ \binom{n}{r}. \tag{2}$$

Equality holds in (2) iff $\mathcal{A} = \emptyset$ or $\mathcal{A} = X^{(r)}$.

Proof. Every set $A \in \mathcal{A}$ contains r elements of $\mathcal{B}$; every set $B \in \mathcal{B} = \partial\mathcal{A}$ is contained in $n - r + 1$ elements of $X^{(r)}$ and so in at most $n - r + 1$ elements of $\mathcal{A}$. Hence

$$|\mathcal{A}|r = \left|\{AB : A \in \mathcal{A}, B \in \mathcal{B}, B \subset A\}\right| \le |\mathcal{B}|(n - r + 1)$$

and so

$$|\mathcal{A}|\binom{n}{r-1} = |\mathcal{A}|\frac{r}{n-r+1}\binom{n}{r} \le |\mathcal{B}|\binom{n}{r},$$

which is equivalent to (2). Now if $\mathcal{A} \ne \emptyset$ and equality holds above then $B \in \mathcal{B}$ implies that all $n - r + 1$ r-sets containing B belong to $\mathcal{A}$. But for any two r-sets $A, A' \in X^{(r)}$ there is a sequence $A_0, B_0, A_1, B_1, A_2, \ldots, B_{l-1}, A_l$ such that $A_i \in X^{(r)}$, $B_i \in X^{(r-1)}$, $B_i = A_i \cap A_{i+1}$ and $A = A_0$, $A' = A_l$ (see Ex. 2). Hence $A \in \mathcal{A}$ implies that $A' \in \mathcal{A}$ and so $\mathcal{A} = X^{(r)}$. ∎

The alert reader may have noticed that a part of the result above is not new to us: inequality (2) in §2 is just Theorem 2 without its uniqueness part; in the proof above we simply reproduced the proof of (2.2). Precisely this inequality was used by Sperner (1928) to prove his theorem.

Corollary 4. *Let* $0 \le s < r \le n$, $\mathcal{A} \subset X^{(r)}$ *and*

$$\mathcal{B} = \{B \in X^{(s)} : B \subset A \quad \text{for some } A \in \mathcal{A}\}.$$

Then

$$|\mathcal{B}|/\binom{n}{s} \ge |\mathcal{A}|/\binom{n}{r} \tag{3}$$

Equality holds in (3) iff $\mathcal{A} = \emptyset$ *or* $\mathcal{A} = X^{(r)}$. ∎

An easy consequence of this corollary is the following extension of the LYM inequality. This extension is a special case of a lemma of Bollobás (1965).

Theorem 5. *Inequality (1) holds, with equality iff* $\mathcal{F} = X^{(k)}$ *for some* k.

Proof. Define $\mathcal{G}_n = \mathcal{F}_n$ and for $k < n$ set $\mathcal{G}_k = \mathcal{F}_k \cup \partial\mathcal{G}_{k+1}$. Put $\varphi_k = f_k\binom{n}{k}^{-1}$ and $\gamma_k = |\mathcal{G}_k|\binom{n}{k}^{-1}$. Note that every element of $\partial\mathcal{G}_{k+1}$ is contained in an element of $\mathcal{F}$. Since $\mathcal{F}$ is a Sperner system, $\mathcal{F}_k$ and $\partial\mathcal{G}_{k+1}$ are disjoint. Therefore, by Theorem 3,

$$\gamma_k \ge \varphi_k + \gamma_{k+1}, \tag{4}$$

with equality if γ_{k+1} is 0 or 1. As $\gamma_n = \varphi_n$, by induction on i we see that

$$\gamma_{n-i} \ge \varphi_{n-i} + \gamma_{n-(i-1)} \ge \varphi_{n-i} + \varphi_{n-i+1} + \dots + \varphi_n.$$

Consequently

$$1 \ge \gamma_0 \ge \sum_{k=0}^{n} \varphi_k = \sum_{k=0}^{n} f_k \binom{n}{k}^{-1}, \tag{5}$$

proving (1).

If equality holds in (5) then equality holds in (4) for every $k < n$, implying that $\mathcal{F} = X^{(k)}$ for some k. ∎

In §4 we shall improve (2) considerably; for every m, $0 \le m \le \binom{n}{r}$, we shall determine the minimum of $|\partial\mathcal{A}|$ as $\mathcal{A}$ runs over all sets satisfying $\mathcal{A} \subset X^{(r)}$ and $|\mathcal{A}| = m$. This improvement will enable us to determine all the sequences $f_0, f_1, \dots, f_n$ for which there is a Sperner family $\mathcal{F}$ satisfying $|\mathcal{F}_k| = |\mathcal{F} \cap X^{(k)}| = f_k$.

To conclude this section, let us note a simple extension of Theorem 2 to certain partially ordered sets. Let $P = (S, <)$ be a poset. A function

$r : S \to \{0, 1, 2, \dots\}$ is a *rank function* on P if $r(b) = r(a) + 1$ whenever b covers a; a poset is *ranked* if it has a rank function. The *level sets* of a ranked poset $(S, <)$ with rank function r are the sets $\{a \in S : r(a) = k\}$. The edges of the covering graph join vertices in neighbouring level sets. The level sets are said to be *regularly connected* if all elements on the kth level dominate the same number of elements on the $(k-1)$st level and are dominated by the same number of elements on the $(k+1)$st level. Note that $r(A) = |A|$ is a rank function on $\mathcal{P}(X)$ and the level sets are $X^{(0)}, X^{(1)}, \dots, X^{(n)}$. Furthermore, $\mathcal{P}(X)$ has regularly connected level sets: every $A \in X^{(k)}$ dominates k elements of $X^{(k-1)}$ and is dominated by $n-k$ elements of $X^{(k+1)}$.

A *chain* in a poset is a set of elements any two of which are comparable and an *antichain* is a set of elements no two of which are comparable. A Sperner family is precisely an antichain in $\mathcal{P}(X)$. Thus the following result is an extension of Theorem 2.

Theorem 6. *Let S be a ranked poset with regularly connected level sets $S_0, S_1, \dots, S_n$ and let $F \subset S$ be an antichain. Set $s_i = |S_i|$ and $f_i = |F \cap S_i|$. Then*

$$\sum_{i=0}^{n} f_i/s_i \le 1. \tag{6}$$

In particular, if no level set has more than s elements then $|F| \le s$.

Proof. Suppose every $x \in S_i$ dominates a_i elements of S_{i-1} and is dominated by b_i elements of S_{i+1}. Every element of S_k is contained in $\left(\prod_{i=1}^{k} a_i\right)\left(\prod_{i=k}^{n-1} b_i\right)$ maximal chains, i.e. chains of the form $x_0 < x_1 < \dots < x_n$, $x_i \in S_i$. Thus if there are m maximal chains in S then $m = s_k\left(\prod_{i=1}^{k} a_i\right)\left(\prod_{i=k}^{n-1} b_i\right)$ for each k, $1 \le k \le n$. Since every chain meets F in at most one element,

$$\sum_{k=0}^{n} f_k \left(\prod_{i=1}^{k} a_i\right)\left(\prod_{i=k}^{n-1} b_i\right) \le m$$

which implies (6). ∎

Just a few more remarks about posets. A beautiful and simple theorem of Dilworth (1950) (see also, e.g., Bollobás (1979, p. 57)) states that the minimum number of chains covering a (finite) poset is precisely the maximal size of an antichain. In this sense it is not by chance that Theorem 1 could be proved by partitioning $\mathcal{P}(X)$ into chains: if $\mathcal{P}(X)$

does not contain more than $m = \binom{n}{\lfloor n/2 \rfloor}$ incomparable elements then we must be able to partition $\mathcal{P}(X)$ into m chains.

Exercises

1. Use Exercise 2.6 to give the following variant of the proof of Theorem 1: $\mathcal{P}(X)$ can be written as the union of $\binom{n}{\lfloor n/2 \rfloor}$ chains of the form $\emptyset \subset A_{i_1} \subset A_{i_2} \subset \ldots \subset A_{i_{n-1}} \subset X$, where $|A_{i_k}| = k$.

2. Let $\emptyset \neq \mathcal{A} \subset X^{(r)}$, $r \geq 1$, such that every set $A \in X^{(r)}$ containing a set $B \in \partial\mathcal{A}$ belongs to $\mathcal{A}$. Prove that $\mathcal{A} = X^{(r)}$.

3. Prove the following simple extension of Theorem 3, namely the local form of Theorem 6. Let S be a ranked poset consisting of two regularly connected level sets: $S = S_1 \cup S_2$. Let ∂A be the set of elements dominated by elements of a set $A \subset S_2$. Prove that

$$|\partial A|/|S_1| \geq |A|/|S_2|.$$

Show also that if the covering graph of S is connected then equality holds iff $A \neq \emptyset$ or $A = S_2$. (A graph is *disconnected* if its vertex set V has a partition into two non-empty sets, say $V = V_1 \cup V_2$, such that there is no edge from V_1 to V_2. A graph which is not disconnected is *connected.*)

4. Let S, s_i, F and f_i be as in Theorem 6. Suppose the restriction of the covering graph to $S_k \cup S_{k-1}$ is connected for every k, $1 \leq k \leq n$. Show that equality holds in (6) iff $F = S_k$ for some k.

5. Let $1 \leq k \leq n/2$ and let $\mathcal{F}$ be a Sperner family consisting of sets of size at most k. $(\mathcal{F} \subset \bigcup_{j=0}^{k} X^{(j)})$. Show that $|\mathcal{F}| \leq \binom{n}{k}$.

6. Let $k \geq 1$ be fixed and let $\mathcal{F}$ be a Sperner family containing at least one set of size at most k, at least one set of size at least $n-k$ and no set whose size is strictly between k and $n-k$. Show that there is a constant c_k such that $|\mathcal{F}| \leq c_k n^{k-1}$. What is the maximum of $|\mathcal{F}|$ for $k = 2$?

7. Prove the following extension of the LYM inequality. Let $\mathcal{F} \subset \mathcal{P}(X)$ be a set system not containing $s+1$ nested sets: $F_1 \subset F_2 \subset \ldots \subset F_{s+1}$, and write $f_k = |\mathcal{F} \cap X^{(k)}|$. Prove that

$$\sum_{k=0}^{n} f_k \binom{n}{k}^{-1} \leq s.$$

In particular, $|\mathcal{F}|$ is at most the sum of the s largest binomial coefficients. (Hint: every maximal chain contains at most s elements of $\mathcal{F}$.)

8. Show that $\mathcal{F}$ satisfies the conditions of Exercise 7 iff it is the union of at most s Sperner systems.

9. Suppose $\mathcal{A} \subset \mathcal{P}(X)$ is an *ideal* (also called a *monotone decreasing set system*), i.e. if $A \in \mathcal{A}$ and $B \subset A$ then $B \in \mathcal{A}$. Use Theorem 3 to show that the average size of a set in $\mathcal{A}$ is at most $n/2$.

Deduce that if $\mathcal{B}$ is a subset of $\mathcal{P}(X)$ is a *monotone increasing set system*, i.e. $B \in \mathcal{B}$ whenever $A \subset B$, then the average size of a set in B is at least $n/2$.

§4. THE LITTLEWOOD-OFFORD PROBLEM

The aim of this brief section is to present Kleitman's beautiful solution to a problem of Littlewood and Offord. When studying the number of real zeros of random polynomials, Littlewood and Offord (1943) arrived at the following problem. Let $z_1, z_2, \ldots, z_n$ be complex numbers of modulus at least 1. Form all 2^n sums of the form $z_{i_1} + z_{i_2} + \ldots + z_{i_t}$, $1 \leq i_1 < i_2 < \ldots < i_t \leq n$. (The value of the empty sum is 0). At most how many of these sums can differ from each other by less than 1? Littlewood and Offord gave a bound which was good enough for their purpose (see also Littlewood (1982, vol. 2, pp. 1333–1344)) but a complete solution was found only considerably later.

Erdős (1945) noticed that for real numbers Sperner's theorem implies a best possible bound. Indeed, suppose $x_1, x_2, \ldots, x_n$ are real numbers of modulus at least 1. For $A \subset [n]$ set $x_A = \sum_{i \in A} x_i$ (by convention, $x_\emptyset = 0$). At most how many of the 2^n sums differ by less than 1 from each other. In this problem x_i may be replaced by $-x_i$ (replace A by $A \triangle \{i\}$ and x by $x - x_i$) so we may assume that $x_i \geq 1$ for every i. Let $\mathcal{F} \subset \mathcal{P}(X)$ be such that $|x_A - x_B| < 1$ for $A, B \in \mathcal{F}$. Then $\mathcal{F}$ is a Sperner system since if $A \subset B \subset [n]$ and $A \neq B$ then $|x_A - x| + |x_B - x| \geq x_B - x_A = x_{B \setminus A} \geq 1$ so at most one of A and B can belong to $\mathcal{F}$. Hence, by Theorem 2.1, $|\mathcal{F}| \leq \binom{n}{\lfloor n/2 \rfloor}$. This bound is clearly best possible: if $x_1 = x_2 = \ldots = x_n = 1$ then $\binom{n}{\lfloor n/2 \rfloor}$ of the sums x_A are $\lfloor n/2 \rfloor$.

Kleitman (1965) and Katona (1966) proved that the bound $\binom{n}{\lfloor n/2 \rfloor}$ holds for sums of complex numbers as well. In fact, Kleitman (1970) proved that, somewhat surprisingly, instead of complex numbers we may even take vectors in an arbitrary normed space. We shall present this latter proof since it is particularly simple and elegant.

The key idea in solving the Littlewood-Offord problem is to adapt a proof of Sperner's theorem to the setting of sums of vectors, rather than simply apply Sperner's theorem. First let us have another look at $\mathcal{P}(X)$. In the proof of Theorem 3.1 we showed that $\mathcal{P}(X)$ can be partitioned into $\binom{n}{\lfloor n/2 \rfloor}$ chains. We shall show now that these chains can be chosen to be of a special form.

Call a chain $A_1 \subset A_2 \subset \ldots \subset A_k$ in $\mathcal{P}(X)$ *symmetric* if $|A_{i+1}| = |A_i| + 1$ and $|A_1| = n - |A_k|$. Thus a chain is symmetric if it goes from a level l to the level $k = n - l$, and has a set on every level between. In particular, every symmetric chain has a set at level $\lfloor n/2 \rfloor$. Can $\mathcal{P}(X)$ be partitioned into symmetric chains? If there is such a partition then, as every symmetric chain meets $X^{(\lfloor n/2 \rfloor)}$ in one set, there must be $\binom{n}{\lfloor n/2 \rfloor}$ non-empty chains in the partition.

It is rather tempting to dismiss this question and claim that the answer is a trivial "yes". To justify this claim, one has the following "solution". Say $n = 2m$ and partition the lower half $\bigcup_{k=0}^{m} X^{(k)}$ into chains, as in the proof of Theorem 2.1, each chain ending in an element of $X^{(m)}$, the middle level. The set $\mathcal{P}(X)$ being symmetric about $X^{(m)}$, the partition of the lower half can be flipped over to produce a partition of the upper half $\bigcup_{k=m}^{n} X^{(k)}$. The two partitions mesh together to produce a symmetric partition of $\mathcal{P}(X)$. Where is the hole in this argument? In the use of the symmetry: the upper half of $\mathcal{P}(X)$ does look like the lower half, but there is no symmetry mapping the lower half into the upper half and keeping the elements of $X^{(m)}$ fixed. Though the argument above is incorrect, the answer to the question is still "yes", but with a not entirely trivial proof.

Theorem 1. *$\mathcal{P}(X)$ can be partitioned into disjoint symmetric chains.*

Proof. Let us apply induction on $n = |X|$. For $n = 1$ there is nothing to prove so assume that $n > 1$ and the assertion holds for smaller values of n. Set $X = [n]$, $Y = [n-1]$ and let $\mathcal{P}(Y) = C_1 \cup C_2 \cup \ldots \cup C_s$ be a partition into non-empty symmetric chains. Note that each chain C_i contains precisely one set of size $\lfloor (n-1)/2 \rfloor$ so $s = \binom{n-1}{\lfloor (n-1)/2 \rfloor}$. Suppose $C_i = \{A_1, A_2, \ldots, A_k\}$, where $A_1 \subset A_2 \subset \ldots \subset A_k$. Set $C_i' = \{A_1, A_2, \ldots, A_k, A_k \cup \{n\}\}$ and $C_i'' = \{A_1 \cup \{n\}, A_2 \cup \{n\}, \ldots, A_{k-1} \cup \{n\}\}$. (For $k = 1$ we have $C_i'' = \emptyset$.) Then C_i' and C_i'' are symmetric chains and clearly

$$\mathcal{P}(X) = \bigcup_{i=1}^{s} C_i' \cup \bigcup_{i=1}^{s} C_i''. \tag{1}$$

Is this a partition? It is indeed. If $A \subset Y$ then only C_i' contains A where C_i is the chain in $\mathcal{P}(Y)$ containing A. If $A = B \cup \{n\}$ where $B \subset Y$

then $B \in C_i$ for some i. If B is the maximal element of C_i then C_i' is the only chain containing A, otherwise A is contained only in C_i''. ∎

A quick reading of the proof above seems to reveal a contradiction. How many chains are there in the partition (1) we constructed? Twice as many as in the partition of $\mathcal{P}(Y)$ we started out with, i.e. $2\binom{n-1}{\lfloor (n-1)/2 \rfloor}$. How many chains do we expect in a symmetric partition of $\mathcal{P}(X)$? Exactly $\binom{n}{\lfloor n/2 \rfloor}$. But $\binom{n}{\lfloor n/2 \rfloor} \neq 2\binom{n-1}{\lfloor (n-1)/2 \rfloor}$ if n is odd! What has gone wrong? Nothing at all, for we have $\binom{n}{\lfloor n/2 \rfloor}$ chains in a partition of $\mathcal{P}(X)$ into non-empty symmetric chains, while in (1) many of our chains may be empty. To be precise, if C_i consists of just one set, which then must be of size $(n-1)/2$, then $C_i'' = \emptyset$, so C_i'' is not included into the partition into non-empty chains. Thus, if $n = 2m$ then we do create $2\binom{2m-1}{m-1} = \binom{2m}{m}$ non-empty chains, but if $n = 2m+1$ then we create only

$$2\binom{2m}{m} - \left\{\binom{2m}{m} - \binom{2m-1}{m-1}\right\} = \binom{2m}{m} + \binom{2m-1}{m-1} = \binom{2m+1}{m}$$

chains, just the right number!

We formulate the main result, Kleitman's (1970) theorem, in terms of normed spaces, though the result does not become less surprising and beautiful if the reader substitutes a Euclidean space $\Re^n$ for B.

Theorem 2. *Let B be a normed space and let $x_1, x_2, \ldots \in B$, $\|x_i\| \geq 1$, $i = 1, 2, \ldots$. Then there are at most $\binom{n}{\lfloor n/2 \rfloor}$ vectors of the form $x_{i_1} + x_{i_2} + \ldots + x_{i_t}$, $1 \leq i_1 < i_2 < \ldots < i_t \leq n$, $t = 0, 1, \ldots, n$, such that the difference of any two of these vectors has norm less than 1.*

Proof. For $A \subset [n]$ set $x_A = \sum_{i \in A} x_i$, just as in the argument about real numbers. Call a set $\mathcal{D} \subset \mathcal{P}([n])$ *sparse* if $A, B \in \mathcal{D}$ and $A \neq B$ imply $\|x_A - x_B\| \geq 1$. Call a partition $\mathcal{P}([n]) = \bigcup_{j=1}^{s} \mathcal{D}_j$ *symmetric* if $|\mathcal{D}_j| \in \{n+1, n-1, n-3, \ldots\} \setminus \{0\}$ and there are precisely $\binom{n}{i} - \binom{n}{i-1}$ sets $\mathcal{D}_j$ of order $n+1-2i$, $i = 0, 1, \ldots, \lfloor n/2 \rfloor$, where $\binom{n}{-1}$ is taken to be 0. Note that we must have $s = \binom{n}{\lfloor n/2 \rfloor}$. Note also that a partition of $\mathcal{P}([n])$ into non-empty symmetric chains is a symmetric partition — in fact, symmetric partitions have been defined by copying the numerical chracteristics of a partition into non-empty symmetric chains.

We shall prove somewhat more than claimed by Theorem 2, namely we shall show that $\mathcal{P}([n])$ has a symmetric partition into sparse sets. The proof of this is very similar to the proof of Theorem 1. We apply induction on n. The case $n = 1$ being trivial, we assume that $n > 1$ and the assertion holds for smaller values of n. Let $\mathcal{P}([n-1]) = \bigcup_{j=1}^{s} \mathcal{D}_j$ be an appropriate symmetric partition. Let f be a support functional at x_n, i.e. let $f \in B^*$, $\|f\| = 1$ and $f(x_n) = \|x_n\| \geq 1$. (If B is the Euclidean space $\Re^n$ then we may assume that $x_n = (\xi, 0, 0, \ldots, 0)$, $\xi \geq 1$, and so f is the first coordinate functional: if $y = (\eta_1, \eta_2, \ldots, \eta_n)$ then $f(y) = \eta_1$.)

Let $\mathcal{D}_j = \{A_1, A_2, \ldots, A_k\}$ and let l be such that

$$f(x_{A_l}) \geq f(x_{A_i})$$

for $1 \leq i \leq k$.

Set $\mathcal{D}_j' = \{A_1, A_2, \ldots, A_k, A_l \cup \{n\}\}$ and $\mathcal{D}_j'' = \{A_1 \cup \{n\}, \ldots, A_{l-1} \cup \{n\}, A_{l+1} \cup \{n\}, \ldots, A_k \cup \{n\}\}$. Then, taking only the non-empty sets $\mathcal{D}_j'$, $\mathcal{D}_j''$, we obtain a symmetric partition of $\mathcal{P}([n])$.

It is clear that $\mathcal{D}_j''$ is sparse. But why is $\mathcal{D}_j'$ sparse? Because for $A_m \in \mathcal{D}_j'$ we have

$$\|x_{A_l \cup \{n\}} - x_{A_m}\| \geq f(x_{A_l \cup \{n\}} - x_{A_m})$$

$$= f(x_n) + f(x_{A_l}) - f(x_{A_m}) \geq f(x_n) \geq 1. \qquad \blacksquare$$

Corollary 3. *For $x \in B$ there are at most $\binom{n}{\lfloor n/2 \rfloor}$ sums of the form $x_{i_1} + x_{i_2} + \ldots + x_{i_t}$, $1 \leq i_1 < i_2 < \ldots < i_t \leq n$, $t = 0, 1, \ldots, n$ at distance less than $\frac{1}{2}$ from x.* ■

Exercises

1. Let $x, x_1, x_2, \ldots, x_n \in \Re^n$, with each x_i having length at least 1, and consider all 2^n sums $\sum_{i=1}^{n} \epsilon_i x_i$ where $\epsilon_i = 1$ or -1. Show that at most $\binom{n}{\lfloor n/2 \rfloor}$ of these sums are at a distance less than 1 from x. (This is the extension of the original form of the Littlewood-Offord lemma — it is equivalent to Theorem 2.)

2. Prove the following extension of Sperner's theorem (Theorem 3.1), due to Katona (1966). Let $X = X_1 \cup X_2$ be a partition of X and let $\mathcal{F} \subset \mathcal{P}(X)$ be a set system such that if $A, B \in \mathcal{F}$, $A \subset B$ and $A \neq B$ then $A \cap X_1 \neq B \cap X_1$ and $A \cap X_2 \neq B \cap X_2$. Then

$$|\mathcal{F}| \leq \binom{n}{\lfloor n/2 \rfloor}.$$

(Assume $|X_1| = n_1 = 2m_1$ and $|X_2| = n_2 = 2m_2$. Partition $\mathcal{P}(X_1)$ into symmetric chains $C_1, \ldots, C_t$, $t = \binom{n_1}{m_1}$, with C_i containing $s_i = 2l_i + 1$ sets. For $A \subset X_1$ let $\mathcal{B}(A) = \{B \subset X_2 : A \cup B \in \mathcal{F}\}$. Note that $\mathcal{B}(A)$ is a Sperner system so, by Exercise 3.7, the system $\bigcup_{A \in C_i} \mathcal{B}(A)$ has at most $c(i) = \sum_{j=-l_i}^{l_i} \binom{n_2}{m_2 - j}$ sets. Check that

$$\sum_{i=1}^{t} c(i) \leq \binom{n}{\lfloor n/2 \rfloor}.$$

Change the argument appropriately if at least one of $|X_1|$ and $|X_2|$ is odd.)

3. Deduce from Exercise 2 the following special case of Exercise 1: if $z_1, z_2, \ldots, z_n$ are complex numbers of modulus at least 1 then at most $\binom{n}{\lfloor n/2 \rfloor}$ sums of the form $\sum_{i=1}^{n} \epsilon_i z_i$, $\epsilon_i = 1$ or -1, are in the interior of a circle of radius 1.

(We may assume that $\operatorname{Re} z_i \geq 0$ for every i. Let $X_1 = \{i : \operatorname{Im} z_i \geq 0\}$ and $X_2 = \{i : \operatorname{Im} z_i < 0\}$. Let $\mathcal{F} = \{A \subset [n] : \sum_{i \in A} z_i - \sum_{i \notin A} z_i$ is in the interior of the circle$\}$. Check that $\mathcal{F}$ satisfies the conditions of Exercise 2.)

4. Take a set of $2k \leq n$ paired parentheses and add to it $n - 2k$ left parentheses in arbitrary places but outside the closed parentheses. Change the leftmost new left parenthesis to a right one, then change the next leftmost to a right one, and so on. In this way we get $n + 1 - 2k$ sequences of parentheses. With each sequence associate the set of the places of the right parentheses. Note that we get a chain of subsets of $X = [n]$, even more, a symmetric chain.

For example, for $n = 9$ and $k = 3$, starting with (( )) ( ) we can insert the new left parentheses before the first, between the fourth and the fifth and after the sixth so, among others, we may take ((( )) (( ) (and obtain the further sequences) (( )) (( ) (,) (( ))) ( ) (and) (( ))) ( )) and so the symmetric chain $458 \subset 1458 \subset 14568 \subset 145689$.

Prove that all these chains give a partition of $\mathcal{P}(X)$ into symmetric chains. Show also that this is exactly the partition defined in the proof of Theorem 1. (de Bruijn, Tengbergen and Kruyswijk (1952), Greene and Kleitman (1976))

5. (Harder) Two partitions of $\mathcal{P}(X)$ into chains are said to be *orthogonal* if no two subsets of X are in the same chain in both partitions.

Prove the theorem of Shearer and Kleitman (1979) stating that for $n \geq 2$ there exist two orthogonal partitions of $\mathcal{P}(X)$ into chains.

(Give explicit constructions for $n = 2$ and 3. For $n \geq 4$ take the partition into symmetric chains given in the proof of Theorem 1 and characterized in Exercise 4. Replacing every set by its complement, obtain another partition. Prove that these partitions are almost orthogonal: only $\emptyset$ and X belong to the same chain in both partitions. Change one of these partitions slightly by transferring $\emptyset$ to a suitable chain to obtain two orthogonal partitions.)

6. Consider a partition of $\mathcal{P}(X)$ into symmetric chains and sum the cardinalities of the largest sets in the chains. Show that the sum is $\frac{n-1}{2}\binom{n}{\lfloor n/2 \rfloor} + 2^{n-1}$.

7. Let $p = 2n$, $X = [n]$, $Z = [p]$, $Y = Z \setminus X = \{n+1, \ldots, p\}$, $m = \binom{n}{\lfloor n/2 \rfloor}$ and let $\mathcal{P}(X) = \sum_{i=1}^m C_i$ and $\mathcal{P}(Y) = \sum_{i=1}^m C_i'$ be partitions into symmetric chains. For a chain $C_i = \{A_k, A_{k+1}, \ldots, A_{n-k}\}$, $A_k \subset A_{k+1} \subset \ldots \subset A_{n-k}$, $|A_j| = j$, let α_i be a sequence *string* of length $n-k$ whose first k terms form A_k and for $j > k$ the jth term is x_j where $A_j \setminus A_{j-1} = \{x_j\}$; define β_i similarly for C_i'. Let $\overline{\alpha}_i$ be the string α_i written in the reverse order and let γ be the following concatenation of strings:

$$\gamma = \overline{\alpha}_1\beta_1\overline{\alpha}_1\beta_2 \ldots \overline{\alpha}_1\beta_m\overline{\alpha}_2\beta_1\overline{\alpha}_2\beta_2 \ldots \overline{\alpha}_2\beta_m \ldots$$

$$\ldots \overline{\alpha}_m\beta_1\overline{\alpha}_m\beta_2 \ldots \overline{\alpha}_m\beta_m.$$

Show that every subset of $Z = [p]$ is the set of some consecutive terms of γ. Make use of the result of Exercise 6 to prove that γ has length

$$(n-1)m^2 + 2^n m \sim \frac{2}{\pi}2^{2n} = \frac{2}{\pi}2^p.$$

(Lipski (1978))

8. For $n = 3$ write out explicitly the string γ in Exercise 7.

§5. SHADOWS

Given a hypergraph $\mathcal{A} \subset X^{(r)}$, the *lower shadow* of $\mathcal{A}$ is

$$\partial_l(\mathcal{A}) = \{B \in X^{(r-1)} : B \subset A \quad \text{for some } A \in \mathcal{A}\},$$

and the *upper shadow* of $\mathcal{A}$ is

$$\partial_u(\mathcal{A}) = \{B \in X^{(r+1)} : B \supset A \quad \text{for some } A \in \mathcal{A}\}.$$

Usually we shall consider only the lower shadow, so when there is no danger of confusion, we shall write ∂ for ∂_l. If $r = 0$ and $\emptyset \neq \mathcal{A} \subset X^{(r)} = X^{(0)} = \{\emptyset\}$ then $\mathcal{A} = \{\emptyset\}$ so $\partial_l(\mathcal{A}) = \emptyset$ and $\partial_u(\mathcal{A}) = X^{(1)}$, and if $r = n$ and $\emptyset \neq \mathcal{A} \subset X^{(r)} = X^{(n)} = \{X\}$ then $\mathcal{A} = \{X\}$ so $\partial_l(\mathcal{A}) = X^{(n-1)}$ and $\partial_u(\mathcal{A}) = \emptyset$. Therefore, when considering shadows of a hypergraph $\mathcal{A}$, we shall assume that $\mathcal{A} \subset X^{(r)}$, $1 \leq r \leq n-1$.

The local LYM inequality (Theorem 3.3) states that

$$|\partial\mathcal{A}| = |\partial_l\mathcal{A}| \geq \frac{|\mathcal{A}|}{\binom{n}{r}}\binom{n}{r-1}. \tag{1}$$

Can we do better than (1)? Even more, what is the greatest lower bound for $|\partial\mathcal{A}|$, the cardinality of the lower shadow of a hypergraph $\mathcal{A} \subset X^{(r)}$, in terms of n, r and $|\mathcal{A}|$? This important problem was solved by Kruskal (1963). The result was rediscovered by Katona in 1968 and, in a more general form, by Clements and Lindström in 1969, and has come to be known as the Kruskal-Katona theorem. Our aim in this section is to prove this fundamental theorem.

Before launching into the theorem, we shall prepare the ground at a leisurely pace. After our preliminary work the proof itself will need very little effort.

Of the several natural orders on $X^{(r)}$, the set of all r-subsets of $X = [n] = \{1, 2, \ldots, n\}$, two tend to be particularly popular. Write $A, B, \ldots \in X^{(r)}$ as $A = \{a_1, a_2, \ldots, a_r\}$, $B = \{b_1, b_2, \ldots, b_r\}, \ldots$ with $a_1 < a_2 < \ldots < a_r$, $b_1 < b_2 < \ldots < b_r, \ldots$. In the *lexicographic* (or simply *lex*) *order* $A < B$ if either $a_1 < b_1$, or $a_1 = b_1$ and $a_2 < b_2$, or $a_1 = b_1$, $a_2 = b_2$ and $a_3 < b_3$, or ... $a_1 = b_1$, $a_2 = b_2, \ldots, a_{r-1} = b_{r-1}$ and $a_r < b_r$. In the *colexicographic* (or *colex*) *order* we have $A < B$ if $A \neq B$ and for $s = \max\{t : a_t \neq b_t\}$ we have $a_s < b_s$. Thus $A < B$ in the colex order if either $a_r < b_r$, or $a_r = b_r$ and $a_{r-1} < b_{r-1}$, or $a_r = b_r$, $a_{r-1} = b_{r-1}$ and $a_{r-2} < b_{r-2}$, or ... $a_r = b_r$, $a_{r-1} = b_{r-1}, \ldots, a_2 = b_2$ and $a_1 < b_1$.

Note that $\{3,4,7,9\} < \{3,5,6,9\}$ in the lex order and $\{3,5,6,9\} < \{3,4,7,9\}$ in the colex order. It is immediate that both lex and colex are indeed *orders* on $X^{(r)}$, i.e. if $A, B, C \in X^{(r)}$ then precisely one of $A = B, A < B$ and $B < A$ holds, and if $A < B$ and $B < C$ then $A < C$. As customary, by $A \leq B$ we mean that $A < B$ or $A = B$.

Throughout the book we shall work with the colex order so an unspecified order on $X^{(r)}$ will always mean the colex order; as we shall see, this order is extremely useful in the study of set systems. Let us explore some properties of the colex order. Just as a matter of curiosity, what is the relationship between lex and colex orders? To get the colex order, take the lex order with respect to the reverse order on $[n] = \{1, 2, \ldots, n\}$, and reverse it. Thus, using the reverse order on [9], in the lex order we have $\{9,7,4,3\} < \{9,6,5,3\}$ so, reversing it, we find that $\{3,5,6,9\} < \{3,4,7,9\}$ in the colex order. Let us list the first 10 elements of $[8]^{(3)}$ in increasing (colex) order: 123, 124, 134, 234, 125, 135, 235, 145, 245, 345. We try to keep the large elements as small as possible and increase them only when we have run out of other options. Thus the 11th, 12th and 13th elements of $[8]^{(3)}$ are 126, 136 and 236. What is the 14th element? We can get away without increasing 6 but we have to increase 3 to 4; having increased 3 we can take 1 instead of 2: 146. What is the set of the first $\binom{m_r}{r}$ elements of $X^{(r)}$? A set $A = \{a_1, \ldots, a_r\}$ with $a_r \leq m_r$ is less than any set $B = \{b_1, \ldots, b_r\}$ with $b_r \geq m_r + 1$. Since there are precisely $\binom{m_r}{r}$ sets $A = \{a_1, \ldots, a_r\}$ with $a_r \leq m_r$, the set of the first $\binom{m_r}{r}$ elements of $X^{(r)}$ is precisely $[m_r]^{(r)}$. What comes after these sets? The r-sets ending in $m_r + 1$ (we are trying to keep the last element small!) with the $(r-1)$-sets of the other elements being the first few sets in the colex order.

To formalize this argument as a theorem, for $\mathcal{A} \subset \mathcal{P}(X)$ and $B \subset X$ set

$$\mathcal{A} + B = \{A \cup B : A \in \mathcal{A}\}$$

and if $B \subset A$ for all $A \in \mathcal{A}$, set

$$\mathcal{A} - B = \{A \setminus B : A \in \mathcal{A}\}.$$

Furthermore, for $0 \le m_s < m_{s+1} < \ldots < m_r$ let

$$\mathcal{B}^{(r)}(m_r, \ldots, m_s) = \bigcup_{j=s}^{r} \left([m_j]^{(j)} + \{m_{j+1}+1, \ldots, m_r+1\}\right) \subset X^{(r)}.$$

The condition $0 \le m_s < m_{s+1} < \ldots < m_r$ guarantees that we define a system of r-sets. Though the definition makes sense for all $0 \le m_s < m_{s+1} < \ldots < m_r$ usually we shall assume that $s \le m_s$ since then the parameters $m_r, m_{r-1}, \ldots, m_s$ are determined by the set system. Set

$$b^{(r)}(m_r, \ldots, m_s) = \sum_{j=s}^{r} \binom{m_j}{j}$$

so that $\left|\mathcal{B}^{(r)}(m_r, \ldots, m_s)\right| = b^{(r)}(m_r, \ldots, m_s)$. Furthermore

$$\mathcal{B}^{(r)}(m_r, \ldots, m_s) = [m_r]^{(r)} \cup \left(\mathcal{B}^{(r-1)}(m_{r-1}, \ldots, m_s) + \{m_r+1\}\right) \quad (1)$$

and so

$$b^{(r)}(m_r, \ldots, m_s) = b^{(r)}(m_r) + b^{(r-1)}(m_{r-1}, \ldots, m_s). \quad (2)$$

When enumerating the first m elements of $X^{(r)}$ in the colex order, the value $n = |X|$ has no effect. Indeed, the first $\binom{n-1}{r}$ elements of $[n]^{(r)}$ form precisely $[n-1]^{(r)}$ and for $m \ge n$ the restriction of the colex order on $[m]^{(r)}$ to $[n]^{(r)}$ is independent of m. Putting this slightly differently, the colex orders on $[r]^{(r)} \subset [r+1]^{(r)} \subset [r+1]^{(r)} \subset \ldots$ are just initial segments of the colex order on $\mathbf{N}^{(r)}$, the set of all r-sets on $\mathbf{N}$. Note that this is false for the lex order. The lex order on $\mathbf{N}^{(3)}$ is 1 2 3, 1 2 4, 1 2 5, 1 2 6, ..., 1 3 4, 1 3 5, 1 3 6, 1 3 7, ..., 2 3 4, 2 3 5, 2 3 6, 2 3 7, ..., etc., a not too useful enumeration of $\mathbf{N}^{(3)}$! In the lex order the $\binom{6}{3} = 20$th element of $[6]^{(3)}$ is $\{4,5,6\}$ but the 20th element of $[25]^{(3)}$ is $\{1,2,22\}$.

By now the alert reader may have realized that the binary representation of natural numbers can be used to define the colex order on $\mathbf{N}^{(r)}$ — better still, on $\mathbf{N}^{(<\omega)}$, the set of all finite subsets of $\mathbf{N}$. Map $\mathbf{N}^{(<\omega)}$ into $\mathbf{N}$ by sending a set $A = \{a_1, \ldots, a_r\}$ into $\varphi(A) = \sum_{i=1}^{r} 2^{a_i}$ (and so the empty set $\emptyset$ into $\varphi(\emptyset) = 1$). The map φ sets up a one to one correspondence between $\mathbf{N}^{(<\omega)}$ and the subset $\{1\} \cup 2\mathbf{N}$ of $\mathbf{N}$; use this correspondence to transfer the order from $\mathbf{N}$ onto $\mathbf{N}^{(<\omega)}$. The restriction of this order on $\mathbf{N}^{(<\omega)}$ to $\mathbf{N}^{(r)}$ is precisely the colex order on $\mathbf{N}^{(r)}$: for $A, B \in \mathbf{N}^{(r)}$ we have $A < B$ iff $\varphi(A) < \varphi(B)$.

Theorem 1. *For $m \in \mathbf{N}$ the set of the first m elements of $\mathbf{N}^{(r)}$ in the colex order is $\mathcal{B}^{(R)}(m_r, \ldots, m_r)$ where the numbers $m_s, m_{s+1}, \ldots, m_r$ are the unique natural numbers such that $s \le m_s < m_{s+1} < \ldots < m_r$ and*

$$m = b^{(r)}(m_r, \ldots, m_s) = \sum_{j=s}^{r} \binom{m_j}{j}. \tag{3}$$

Proof. Let us see first by induction on m that $m_r, m_{r-1}, \ldots, m_s$ exist and are determined by m. This is certainly true for $m = 1$: we must have $m_r = r$ and $s = r$. Suppose then that $m \ge 2$. Since for $k \ge r$ we have

$$\sum_{l=0}^{r-1} \binom{k-l}{r-l} = \binom{k+1}{r} - 1 < \binom{k+1}{r},$$

we must have

$$m_r = \max\{k : \binom{k}{r} \le m\}.$$

But then either $m = b^{(r)}(m_r) = \binom{m_r}{r}$ or else there are unique integers $s \le m_s < m_{s+1} < \ldots < m_{r-1}$ such that

$$m - b^{(r)}(m_r) = b^{(r-1)}(m_{r-1}, m_{r-2}, \ldots, m_s)$$

and so, by (2),

$$m = b^{(r)}(m_r, \ldots, m_s),$$

with $m_r, m_{r-1}, \ldots, m_s$ uniquely determined by m.

The argument above very nearly proves that $\mathcal{B}^{(r)}(m_r, \ldots, m_s)$ is the set of the first n elements of $\mathbf{N}^{(r)}$. The set of the first $b^{(r)}(m_r) \le m$ elements is precisely $[m_r]^{(r)} = \mathcal{B}^{(r)}(m_r)$ and the next $m' = m - b^{(r)}(m_r) = b^{(r-1)}(m_{r-1}, \ldots, m_s)$ elements all end in $m_r + 1$ and start in the first m' elements on $\mathbf{N}^{(r-1)}$. Hence, by induction on m (or on r, for that matter), these m' elements form $\mathcal{B}^{(r-1)}(m_{r-1}, \ldots, m_s) + \{m_r + 1\}$. Relation (1) shows that the set of the first m elements of $\mathbf{N}^{(r)}$ is $\mathcal{B}^{(r)}(m_r, \ldots, m_s)$. ∎

What is the mth element of $\mathbf{N}^{(r)}$? If $m = b^{(r)}(m_r, \ldots, m_s)$, as in (3), then it is just the last element of $\mathcal{B}^{(r)}(m_r, \ldots, m_s)$, so it is the last element of $[m_s]^{(s)}$ together with $m_{s+1} + 1, m_{s+2} + 1, \ldots, m_r + 1$:

$\{m_s - s+1, m_s - s+2, \ldots, m_s, m_{s+1}+1, m_{s+2}+1, \ldots, m_r+1\}$. Conversely, let $A = \{a_1, a_2, \ldots, a_r\} \in \mathbb{N}^{(r)}$ with $a_1 < a_2 < \ldots < a_r$. Define $s = \max\{t : 1 \le t \le r, a_t = a_{t-1} + 1\}$. Then

$$\{B \in \mathbb{N}^{(r)} : B \le A\} = \mathcal{B}^{(r)}(a_r - 1, a_{r-1} - 1, \ldots, a_{s+1} - 1, a_s).$$

For example, what is the 1000th element of $\mathbb{N}^{(5)}$? We need $1000 = b^{(5)}(m_5, \ldots)$. Now m_5 is the largest natural number satisfying $\binom{m_5}{5} \le 1000$ so $m_5 = 12$. As $1000 - \binom{12}{5} = 208 > 0$, m_4 is the largest natural number satisfying $\binom{m_4}{4} \le 208$ so $m_4 = 9$. Then we get $208 - \binom{9}{4} = 82$, giving $m_3 = 8$ and $82 - \binom{8}{3} = 26$. From this, $m_2 = 7$ and $26 - \binom{7}{2} = 5$. Finally, $m_1 = 5$. Therefore the first 1000 elements form $\mathcal{B}^{(5)}(12, 9, 8, 7, 5)$ and the 1000th element is $\{5, 8, 9, 10, 13\}$. What are then the 1001st, 1002nd and 1003rd elements? They are $\{6, 8, 9, 10, 13\}$, $\{7, 8, 9, 10, 13\}$ and $\{1, 2, 3, 11, 13\}$. Finally, what are the sets of the first 1001, 1002 and 1003 elements? By applying our rules, we see that they are $\mathcal{B}^{(5)}(12, 9, 8, 7, 6)$, $\mathcal{B}^{(5)}(12, 10)$ and $\mathcal{B}^{(5)}(12, 10, 3)$. Let us note another recursion formula for $b^{(r)}(m_r, \ldots, m_s)$:

$$\begin{aligned} b^{(r)}(m_r, \ldots, m_s) &= b^{(r)}(m_r - 1, \ldots, m_s - 1) \\ &\quad + b^{(r-1)}(m_r - 1, \ldots, m_s - 1) \end{aligned} \tag{4}$$

This relation is immediate from

$$\binom{m_j - 1}{j} + \binom{m_j - 1}{j - 1} = \binom{m_j}{j},$$

but a more combinatorial argument goes as follows. For $\mathcal{A} \subset X^{(r)}$ set

$$\mathcal{A}_0 = \{A \in \mathcal{A} : 1 \notin A\}$$

and

$$\mathcal{A}_1 = \{A \in \mathcal{A} : 1 \in A\}.$$

Then $\mathcal{A}$ is partitioned as $\mathcal{A}_0 \cup \mathcal{A}_1$ and so

$$|\mathcal{A}| = |\mathcal{A}_0| + |\mathcal{A}_1| = |\mathcal{A}_0| + |\mathcal{A}_1 - \{1\}|. \tag{5}$$

Now if $\mathcal{A} = \mathcal{B}^{(r)}(m_r, \ldots, m_s)$ then (5) turns into (4). Thus (4) is just a natural consequence of a natural partition of $\mathcal{B}^{(r)}(m_r, \ldots, m_s)$.

Though we could dwell quite a bit longer on the fascinating properties of the colex order, it is time to consider shadows. What is the (lower) shadow of $\mathcal{B}^{(r)}(m_r,\ldots,m_s)$, where $s \leq m_s < m_{s+1} < \ldots < m_r$? Clearly

$$\partial\mathcal{B}^{(r)}(m_r,\ldots,m_s) = \mathcal{B}^{(r-1)}(m_r,\ldots,m_s).$$

This shows that the shadow of a system of $b^{(r)}(m_r,\ldots,m_s)$ r-sets need not have more than $b^{(r-1)}(m_r,\ldots,m_s)$ elements. The celebrated Kruskal-Katona theorem claims that this example is best possible: the shadow of a system of $b^{(r)}(m_r,\ldots,m_s)$ r-sets (with $s \leq m_s < m_{s+1} < \ldots < m_r$) contains at least $b^{(r-1)}(m_r,\ldots,m_s)$ $(r-1)$-sets. The original proofs have been greatly simplified over the years: thanks to Kleitman (1966b), Daykin, Godfrey and Hilton (1974), Daykin (1974) and Frankl (1984), there is a very elegant and simple proof, based on properties of the colex order and the compression operators $\tilde{R}_{ij}$.

For $i,j \in X$, $i \neq j$, and $A \subset X$ define

$$R_{ij}(A) = \begin{cases} (A\setminus\{j\})\cup\{i\} & \text{if } i \notin A,\ j \in A \\ A & \text{otherwise.} \end{cases}$$

Thus R_{ij} replaces the element j by the element i whenever possible. By definition $|R_{ij}(A)| = |A|$ and if $i < j$ then $R_{ij}(A) \leq A$ in the colex order. Note that R_{ij} is not a $1-1$ map: for $B \subset X\setminus\{i,j\}$ one has $R_{ij}(B\cup\{i\}) = R_{ij}(B\cup\{j\}) = B\cup\{i\}$. To see the action of R_{ij}, let

$$\mathcal{P}_{ij}(X) = \{A \subset X : i \in A,\ j \notin A\}.$$

Then R_{ij} gives a $1-1$ correspondence between $\mathcal{P}_{ji}(X)$ and $\mathcal{P}_{ij}(X)$, and on $\mathcal{P}(X)\setminus\mathcal{P}_{ji}(X)$ the operator R_{ij} is simply the identity. Let us associate with R_{ij} a map $\tilde{R}_{ij}$ sending a set system into another set system: for $\mathcal{A} \subset \mathcal{P}(X)$ let

$$\tilde{R}_{ij}(\mathcal{A}) = \{R_{ij}(A) : A \in \mathcal{A}\} \cup \{A : A, R_{ij}(A) \in \mathcal{A}\}.$$

We call $\tilde{R}_{ij}$ a *compression operator*. Clearly

$$|\tilde{R}_{ij}(\mathcal{A})| = |\mathcal{A}|$$

for all $\mathcal{A} \subset \mathcal{P}(X)$, and

$$\tilde{R}_{ij}(\mathcal{A}) \subset X^{(r)} \quad \text{if} \quad \mathcal{A} \subset X^{(r)}.$$

Call a set system $\mathcal{A} \subset \mathcal{P}(X)$ *left compressed* or simply *compressed* if $\tilde{R}_{ij}(\mathcal{A}) = \mathcal{A}$ whenever $1 \leq i < j \leq n$. Thus $\mathcal{A}$ is left compressed if $A \in \mathcal{A}$,

$j \in A$, $i \notin A$ and $i < j$ imply $R_{ij}(A) = (A \setminus \{j\}) \cup \{i\} \in \mathcal{A}$. Equivalently, $\mathcal{A}$ is left compressed if whenever $A_1 = \{a_s, a_{s+1}, \dots, a_t\} \subset A$, $a_s < a_{s+1} < \dots < a_t$, $B_1 = \{b_s, b_{s+1}, \dots, b_t\} \subset X \setminus A$, $b_s < b_{s+1} < \dots < b_t$ and $b_s < a_s$, $b_{s+1} < a_{s+1}, \dots, b_t < a_t$, the set $(A \setminus A_1) \cup B_1$ also belongs to $\mathcal{A}$.

At this stage one cannot help wondering whether a compressed set of m r-sets is not simply the set of the first m r-sets in the colex order. In fact, this cannot be true for every m because $R_{ij}(A) \leq A$ for $i < j$ not only in the colex order but also in the lex order. For example, $\mathcal{A} = \{1\,2\,3,\ 1\,2\,4,\ 1\,2\,5,\ 1\,2\,6\}$ is compressed but it is rather far from $\mathcal{B}^{(3)}(4) = \{1\,2\,3,\ 1\,2\,4,\ 1\,3\,4,\ 2\,3\,4\}$, the set of the first 4 elements in the colex order.

For us the most important property of a compression operator is that it does not increase a shadow of the system.

Lemma 2. *(i) For $\mathcal{A} \subset X^{(r)}$ and $1 \leq i < j \leq n$ we have*

$$|\partial \mathcal{A}| \geq |\partial \tilde{R}_{ij}(\mathcal{A})|. \tag{6}$$

(ii) For all $\mathcal{A} \subset X^{(r)}$ there is a left compressed set system $\mathcal{A}' \subset X^{(r)}$ such that

$$|\mathcal{A}| = |\mathcal{A}'| \quad \text{and} \quad |\partial \mathcal{A}| \geq |\partial \mathcal{A}'|. \tag{7}$$

Proof. (i) Let $B \in \partial \tilde{R}_{ij}(\mathcal{A}) \setminus \partial \mathcal{A}$. Then $B \in \mathcal{P}_{ij}(X)$ and $R_{ji}(B) \in \partial \mathcal{A} \setminus \partial \tilde{R}_{ij}(\mathcal{A})$. Since $R_{ji} : \mathcal{P}_{ij}(X) \to \mathcal{P}_{ji}(X)$ is $1-1$,

$$|\partial \tilde{R}_{ij}(\mathcal{A}) \setminus \partial \mathcal{A}| = |R_{ji}(\partial \tilde{R}_{ij}(\mathcal{A}) \setminus \partial \mathcal{A})| \leq |\partial \mathcal{A} \setminus \partial \tilde{R}_{ij}(\mathcal{A})|$$

so (6) does hold.

In words: (6) holds because every set B in $\partial \tilde{R}_{ij}(\mathcal{A}) \setminus \partial \mathcal{A}$ contains i but not j; replacing i by j we obtain a set $R_{ji}(B) \in \partial \mathcal{A} \setminus \partial \tilde{R}_{ij}(\mathcal{A})$.

(ii) Let us construct a sequence of systems $\mathcal{A}_0, \mathcal{A}_1, \dots$ as follows. Set $\mathcal{A}_0 = \mathcal{A}$. Suppose we have constructed $\mathcal{A}_0, \mathcal{A}_1, \dots \mathcal{A}_k$. If $\mathcal{A}_k$ is compressed then stop the sequence, otherwise pick an operator $\tilde{R}_{ij}$, $1 \leq i < j \leq n$, for which $\tilde{R}_{ij}(\mathcal{A}_k) \neq \mathcal{A}_k$ and set $\mathcal{A}_{k+1} = \tilde{R}_{ij}(\mathcal{A}_k)$. This sequence has to end in some system $\mathcal{A}_l$ since with

$$w(\mathcal{A}_i) = \sum_{A \in \mathcal{A}_i} \sum_{a \in A} a$$

we have $w(\mathcal{A}_0) > w(\mathcal{A}_1) > \dots$. The system $\mathcal{A}' = \mathcal{A}_l$ is compressed and $|\mathcal{A}'| = |\mathcal{A}|$.

By (6) the cardinalities of the shadows are non-increasing since $\mathcal{A}_{k+1} = \tilde{R}_{ij}(\mathcal{A}_k)$:

$$|\partial \mathcal{A}_0| \geq |\partial \mathcal{A}_1| \geq \ldots \geq |\partial \mathcal{A}_l| = |\partial \mathcal{A}'|,$$

proving (7). ∎

We are about to state and prove the Kruskal-Katona theorem. First let us define a function $\partial^{(r)} : \mathbf{N} \to \mathbf{N}$ as follows: for $m \geq 1$ take the unique representation of m guaranteed by Theorem 1:

$$m = b^{(r)}(m_r, \ldots, m_s)$$

and define

$$\partial^{(r)}(m) = b^{(r-1)}(m_r, \ldots, m_s). \tag{8}$$

Thus $\partial^{(r)}(m)$ is the number of $(r-1)$-sets in the shadow of the first m r-sets in the colex order.

Theorem 3. *Let $r \geq 1$ and $\mathcal{A} \subset X^{(r)}$. Then*

$$|\partial \mathcal{A}| \geq \partial^{(r)}(|\mathcal{A}|) \tag{9}$$

i.e. the shadow of $\mathcal{A}$ is at least as large as the shadow of the first $|\mathcal{A}|$ r-sets in the colex order. If $|\mathcal{A}| = \binom{m_r}{r}$ for some $m_r \geq r$ then equality holds in (9) iff $\mathcal{A} \simeq [m_r]^{(r)}$.

Proof. In proving (9), by Lemma 2(i) we may assume that $\mathcal{A}$ is compressed. If we could say that $\mathcal{A}$ is just the set of the first $|\mathcal{A}|$ elements then (9) would be proved. As it is, we have to work a little to prove (9).

We shall apply a double induction: first on r and then on $m = |\mathcal{A}|$, and we shall use the analogue of the splitting in (5) to deduce (9). If either $r = 1$ or $m = 1$ then the inequality is trivial so suppose $r \geq 2$ and $m \geq 2$ and the inequality holds for $r - 1$ and all values of m and for r and $1, 2, \ldots, m - 1$. Let

$$m = b^{(r)}(m_r, \ldots, m_s)$$

where $s \leq m_s < m_{s+1} < \ldots < m_r$ and define

$$\mathcal{A}_0 = \{A \in \mathcal{A} : 1 \notin A\}$$

and

$$\mathcal{A}_1 = \{A \in \mathcal{A} : 1 \in \mathcal{A}\} = \mathcal{A} \setminus \mathcal{A}_0.$$

Then

$$\partial\mathcal{A}_0 \subset \mathcal{A} - \{1\} \tag{10}$$

since if $B \in \partial\mathcal{A}_0$ then $B \cup \{j\} \in \mathcal{A}_0$ for some $j > 1$ and so $R_{1j}(B \cup \{j\}) = B \cup \{1\} \in \mathcal{A}_1$ and $B \in \mathcal{A}_1 - \{1\}$. By (10),

$$|\partial\mathcal{A}_0| \leq |\mathcal{A}_1 - \{1\}| = |\mathcal{A}_1| : \tag{11}$$

If $j \in A \in \mathcal{A}_1$ and $j > 1$ then

$$A \setminus \{j\} = \big((A \setminus \{1\}) \setminus \{j\}\big) \cup \{1\}$$

so $\partial\mathcal{A}_1$ has the following partition:

$$\partial\mathcal{A}_1 = (\mathcal{A}_1 - \{1\}) \cup \big(\partial(\mathcal{A}_1 - \{1\}) + \{1\}\big). \tag{12}$$

If $\mathcal{A}$ is $\mathcal{B}^{(r)}(m_r, \ldots, m_s)$, namely the set system we hope is extremal, then $|\mathcal{A}_1| = b^{(r-1)}(m_r - 1, \ldots, m_s - 1)$ and $|\mathcal{A}_0| = b^{(r)}(m_r - 1, \ldots, m_s - 1)$, i.e. we get the splitting corresponding to (4). Thus it is natural to distinguish two cases according to the relationship between $|\mathcal{A}_1|$ and $b^{(r-1)}(m_r - 1, \ldots, m_s - 1)$.

Suppose that $|\mathcal{A}_1| < b^{(r-1)}(m_r - 1, \ldots, m_s - 1)$. Then, by (4),

$$|\mathcal{A}_0| > b^{(r)}(m_r - 1, \ldots, m_s - 1)$$

so, by our induction hypothesis,

$$|\partial\mathcal{A}_0| \geq b^{(r-1)}(m_r - 1, \ldots, m_s - 1),$$

contradicting (11). Therefore $|\mathcal{A}_1| \geq b^{(r-1)}(m_r - 1, \ldots, m_s - 1)$.

By (12) and the induction hypothesis applied to $\partial(\mathcal{A}_1 - \{1\}) \subset X^{(r-1)}$, we find that

$$\begin{aligned} |\partial\mathcal{A}_1| &\geq |\mathcal{A}_1| + \partial^{(r-1)}(|\mathcal{A}_1|) \\ &\geq b^{(r-1)}(m_r - 1, \ldots, m_s - 1) + b^{(r-2)}(m_r - 1, \ldots, m_s - 1) \\ &= b^{(r-1)}(m_r, \ldots, m_s), \end{aligned}$$

where in the last step we used (4). Hence

$$|\partial\mathcal{A}| \geq |\partial\mathcal{A}_1| \geq b^{(r-1)}(m_r, \ldots, m_s) = \partial^{(r)}(|\mathcal{A}|),$$

proving (9).

The second assertion is easily read out of the proof above. If we have equality in (9) for some $\mathcal{A}$ with $|\mathcal{A}| = m = \binom{m_r}{r}$, $m_r > r$, then $|\mathcal{A}_1| = b^{(r-1)}(m_r - 1)$ and $|\partial(\mathcal{A}_1 - \{1\})| = b^{(r-2)}(m_r - 1)$. Therefore, by the induction hypothesis applied to $\mathcal{A}_1 - \{1\}$, we see that $\mathcal{A}_1$ is the set of all $\binom{m_r-1}{r-1}$ r-sets containing 1 and contained in $[m_r]$. Then the shadow of $\mathcal{A}_1$ is $[m_r]^{(r-1)}$ and as $\partial\mathcal{A} = \partial\mathcal{A}_1$, we must have $\mathcal{A} = [m_r]^{(r)}$.

It seems that we are done, but not quite, since $\mathcal{A}$ is not our original set system but one obtained from the original by applying a sequence of compression operators. However, to overcome this difficulty, all we have to note is that if $\tilde{R}_{ij}(\mathcal{A}) = [Y]^{(r)}$ for some $Y \subset X$, $|Y| = m_r$, $i \in Y$ and $j \notin Y$, and $\mathcal{A} \neq [Y]^{(r)}$ then $\mathcal{A} = \tilde{R}_{ji}([Y]^{(r)}) = [R_{ji}(Y)]^{(r)}$. Hence $[m_r]^{(r)}$ was obtained from some system $[Y_1]^{(r)}$, that arose from some system $[Y_2]^{(r)}$, etc., where $|Y_1| = |Y_2| = \ldots = m_r$. Thus the system we started out with has to be $[Y]^{(r)}$ for some $Y \subset X$, $|Y| = m_r$, i.e. it is isomorphic to $[m_r]^{(r)} = \mathcal{B}^{(r)}(m_r)$, as claimed. ■

According to the Kruskal-Katona theorem, the shadow of m r-sets cannot be smaller than the shadow of the first m r-sets in the colex order. In addition, the shadow of the first m r-sets in the colex order consists of precisely the first $\partial^{(r)}(m)$ $(r-1)$-sets in the colex order. This very fortunate phenomenon enables one to extend the LYM inequality (Theorem 3.2) to a characterization of the parameters of a Sperner family. This characterization was noted by Clements (1973) and by Daykin, Godfrey and Hilton (1974).

Theorem 4. *Let $f_0, f_1, \ldots, f_n$ be a sequence of non-negative integers. There is a Sperner family $\mathcal{F} \subset \mathcal{P}(X)$ such that $f_i = |\mathcal{F}_i| = |\mathcal{F} \cap X^{(i)}|$, $i = 0, 1, \ldots, n$, iff*

$$g_n = f_n \leq \binom{n}{n}$$

$$g_{n-1} = \partial^{(n)}(g_n) + f_{n-1} \leq \binom{n}{n-1},$$

$$g_{n-2} = \partial^{(n-1)}(g_{n-1}) + f_{n-2} \leq \binom{n}{n-2},$$

$$\vdots$$

$$g_0 = \partial^1(g_1) + f_0 \leq \binom{n}{0}.$$

Proof. Suppose $\mathcal{F}$ is a Sperner family and $f_i = |\mathcal{F}_i|$, $i = 1, 2, \ldots, n$. Define $\mathcal{H}_n \subset X^{(n)}, \mathcal{H}_{n-1} \subset X^{(n-1)}, \ldots$ by setting $\mathcal{H}_n = \mathcal{F}_n$ and

$$\mathcal{H}_j = \partial(\mathcal{H}_{j+1}) \cup \mathcal{F}_j, \qquad 0 \le j \le n-1.$$

Note that

$$\mathcal{H}_{j+1} = \partial^{n-(j+1)} \mathcal{F}_n \cup \partial^{(n-1)-(j+1)} \mathcal{F}_{n-1} \cup \ldots \cup \mathcal{F}_{j+1}$$

so

$$\partial(\mathcal{H}_{j+1}) \cap \mathcal{F}_j = \emptyset$$

and

$$|\partial(\mathcal{H}_{j+1})| + |\mathcal{F}_j| \le \binom{n}{j}.$$

Furthermore, with $h_j = |\mathcal{H}_j|$, by the Kruskal-Katona theorem,

$$h_n = g_n$$

and

$$h_j \ge \partial^{(j+1)}(h_{j+1}) + f_j, \qquad 0 \le j \le n-1.$$

Hence

$$g_j \le h_j \le \binom{n}{j}$$

for $j = n, n-1, \ldots, 0$, as claimed.

Conversely, suppose $g_0, g_1, \ldots, g_n$ satisfy the conditions. Denote by $\mathcal{C}^{(r)}(m)$ the set of the first m elements of $X^{(r)}$ in the colex order and set

$$\mathcal{G}_n = \mathcal{C}^{(n)}(g_n)$$

and

$$\mathcal{G}_j = \mathcal{C}^{(j)}(g_j) \setminus \mathcal{C}^{(j)}(\partial^{(j+1)}(g_{j+1})), \qquad 0 \le j \le n-1.$$

Since

$$\partial(\mathcal{C}^{(j+1)}(g_{j+1})) = \mathcal{C}^{(j)}(\partial^{(j+1)}(g_{j+1})),$$

the family $\mathcal{G} = \bigcup_{j=0}^{n} \mathcal{G}_j$ is a Sperner family, $|\mathcal{G}_n| = g_n = f_n$ and $|\mathcal{G}_j| = g_j - \partial^{(j+1)}(g_{j+1}) = f_j$, $0 \le j \le n-1$. ∎

To conclude this section, let us see how Theorem 3 can be used to give a best possible lower bound for the upper shadow of a set system $\mathcal{A} \subset X^{(r)}$.

Theorem 5. *Let $1 \le r \le n-1$, $\mathcal{A} \subset X^{(r)}$ and let $\mathcal{B}$ be the set of the last $|\mathcal{A}|$ elements of $X^{(r)}$ in the colex order. Then*

$$|\partial_u \mathcal{A}| \ge |\partial_u \mathcal{B}|$$

Proof. For $\mathcal{F} \subset \mathcal{P}(X)$ let $\mathcal{F}^c = \{F^c : F \in \mathcal{F}\}$ be the family of complements of sets in $\mathcal{F}$. Then $\partial_u \mathcal{A} = \left(\partial_l(\mathcal{A}^c)\right)^c$ so

$$|\partial_u \mathcal{A}| = |\partial_l(\mathcal{A}^c)| \ge |\partial_l(\mathcal{C})| = |\partial_u(\mathcal{C}^c)|$$

where $\mathcal{C}$ is the set of the first $|\mathcal{A}|$ elements of $X^{(n-r)}$ in the colex order. Note that $\mathcal{C}^c$ is precisely the set of the last $|\mathcal{A}|$ elements of $X^{(r)}$ in the colex order so $\mathcal{C}^c = \mathcal{B}$. ∎

Exercises

1. What is the value of $b^{(4)}(14,6,5,2)$? Describe the set $\mathcal{B}^{(4)}(14,6,4,2)$!

2. What is the 1001st element of $\mathbf{N}^{(4)}$? What are the 999th, 1000th, 1002nd, 1003rd and 1004th elements?

3. Check that if $\mathcal{A} \subset \mathbf{N}^{(2)}$, $|\mathcal{A}| = 4$ then $|\partial \mathcal{A}| \ge 4$. How many non-isomorphic systems $\mathcal{A}$ are there with $|\partial \mathcal{A}| = 4$?

4. Show that if $m = b^{(r)}(m_r, \ldots, m_2, 1)$ where $r \ge 2$ and $m_2 \ge 3$ then there are several non-isomorphic families $\mathcal{A} \subset \mathbf{N}^{(r)}$ such that $|\mathcal{A}| = m$ and $|\partial \mathcal{A}|$ is as small as possible, namely $b^{(r-1)}(m_r, \ldots, m_2)$.

5. Prove that if $r \ge 3$, $\mathcal{A} \subset \mathbf{N}^{(r)}$, $|\mathcal{A}| = b^{(r)}(m_r, m_{r-1})$ where $r \le m_{r-1} < m_r$, and $|\partial \mathcal{A}| \le b^{(r-1)}(m_r, m_{r-1})$ then $\mathcal{A} \cong \mathcal{B}^{(r)}(m_r, m_{r-1})$.

6. Deduce from Theorem 4 the following result of Daykin, Godfrey and Hilton (1974). If there is a Sperner system $\mathcal{F}$ with parameters $f_i = |\mathcal{F}_i| = |\mathcal{F} \cap X^{(i)}|$ then there is a Sperner system $\mathcal{F}'$ with parameters f_i' such that $f_i' = 0$ if $i > n/2$, $f_{n/2}' = f_{n/2}$ and $f_i' = f_i + f_{n-i}$ for $i < n$.

§6. RANDOM SETS

This section touches on a rather large subject, the theory of random graphs. We shall hardly do more than define some of the terms, prove a consequence of the Kruskal-Katona theorem and note another result about random sets. When talking about random graphs, we shall assume that the reader has encountered the basic concepts of graph theory like connectedness, complete graph, cycle, path and diameter. The reader unfamiliar with these concepts should just skip the remarks about graphs and pass on to random sets. As we wish to keep our convention that the ground set X has n elements, our notation concerning random graphs will be unconventional. For an extensive account of the theory of random graphs the reader should consult Bollobás (1985).

Let $V = \{x_1, x_2, \ldots, x_t\}$ and let $\mathcal{G}^t$ be the set of all graphs on V. Consider the subset $\mathcal{G}(t,k)$ of $\mathcal{G}^t$ consisting of graphs with $k = k(t)$ edges. Setting $n = \binom{t}{2}$ we see that $\mathcal{G}(t,k)$ contains precisely $\binom{n}{k}$ graphs. Turn $\mathcal{G}(t,k)$ into a *probability space* by giving all its elements the same probability, and write $G_{t,k}$ for a random element of $\mathcal{G}(t,k)$; we call $G_{t,k}$ a *random graph of order t and size k*. The *probability* in the space $\mathcal{G}(t,k)$ is denoted by $P_{t,k}$. A *property* Q of graphs is naturally identified with the set of graphs on V having Q. Then $Q_k = Q \cap \mathcal{G}(t,k)$ is the set of graphs in $\mathcal{G}(t,k)$ having property Q. Given a property Q of graphs, $P_{t,k}(Q) = P_{t,k}(Q_k) = |Q_k|/\binom{n}{k}$ is the probability that $G_{t,k}$ has Q. Most graph properties one considers are *monotone*: if $G, H \in \mathcal{G}^t$ and $G \in Q$ then either $H \supset G$ implies that $H \in Q$ (and the property Q is then *monotone increasing*) or else $H \subset G$ implies $H \in Q$ (and the property Q is then *monotone decreasing*). A property Q is monotone increasing iff its *negation*, $\neg Q = Q^c = \mathcal{G} \setminus Q$, is monotone decreasing. The properties of containing a triangle, being connected, having diameter at most d, containing a Hamiltonian cycle and having chromatic number at least s are all monotone increasing properties. It is easy to see (and intuitively obvious) that if Q is a monotone increasing property then $P_{t,k}(Q)$ is a monotone increasing function of k: if we pick more edges then we are more likely to end up with a graph having Q. The theory of random graphs was founded by Erdős and Rényi in the late fifties and early sixties, and by now there is a well developed subject with an immense literature (see Bollobás (1985)). In the theory of random graphs one studies the function $P_{t,k}(Q)$ for various properties Q as $t \to \infty$ and one is particularly interested in the functions $k(t)$ for which $\lim_{t\to\infty} P_{t,k(t)}(Q)$ is 0 or 1. One of the major discoveries of Erdős and Rényi was that many monotone increasing properties appear rather suddenly: the limit above changes from 0 to 1 as $k(t)$ is increased a little; in other words, Erdős and Rényi proved that many monotone increasing properties have so-called threshold functions. We say that *almost every* (a.e.) *random graph* $G_{t,k}$ has Q or that $G_{t,k}$ has Q *almost surely* (a.s.) if $\lim_{t\to\infty} P_{t,k}(Q) = 1$; also we say that *almost no* $G_{t,k}$ *has* Q or that $G_{t,k}$ *fails to have* Q *almost surely* if $\lim_{t\to\infty} P_{t,k}(Q) = 0$.

A function $k^*(t)$ is a *threshold function* for a monotone increasing property Q if whenever $k/k^* \to 0$ almost no $G_{t,k}$ has Q and whenever $k/k^* \to \infty$ almost every $G_{t,k}$ has Q. Erdős and Rényi proved, among others, that connectedness has threshold function $k^*(t) = t\log t$, the property of containing a triangle has threshold function $k^*(t) = t$, the property of containing a complete graph of order 5 has threshold function $t^{3/2}$ and the property of having diameter at most 3 has threshold function $t^{4/3}$.

A random graph $G_{t,k}$ is an example of a *random set*. Let X and $\mathcal{P}(X)$ be as before, and call a set system Q a *property* of the subsets of X. Thus $Q \subset \mathcal{P}(X)$ and the statement "a set $Y \subset X$ has Q" means simply that $Y \in Q$. We call Q *monotone increasing* if $A \in Q$ and $A \subset B \subset X$ imply $B \in Q$; also, Q is *monotone decreasing* or an *ideal* if $A \in Q$ and $B \subset A$ imply $B \in Q$. Once again, Q is monotone increasing iff its negation (complement), $\neg Q = Q^c = \mathcal{P}(X) \setminus Q$, is monotone decreasing.

Turn $X^{(k)}$ into a probability space by taking its elements equiprobable; then a random k-set is just an element of $X^{(k)}$. The *probability that a random k-set of X has Q* is defined to be

$$P_{n,k}(Q) = P_{n,k}(Q_k) = |Q_k|/|X^{(k)}| = |Q_k|/\binom{n}{k}$$

where $Q_k = Q \cap X^{(k)}$ is the kth level set of Q.

The terms "almost every", "almost no", "almost surely" are defined as before and so is a *threshold function*. For example, let Q be the property that the set contains two consecutive elements of X, i.e. it contains a set $\{i, i+1\}$ where $1 \le i \le n-1$. Then $k^*(n) = n^{1/2}$ is a threshold function for Q (see Ex. 3).

To see that a random graph is just a special random set, identify the set $V^{(2)}$ of all $n = \binom{t}{2}$ pairs of vertices with the set X. The edges of a graph in $\mathcal{G}^t$ are chosen from $V^{(2)}$. Then $\mathcal{G}^t$ is identified with $\mathcal{P}(X)$ and $\mathcal{G}(t,k)$ with $X^{(k)}$; a random $G_{t,k}$ corresponds to a random k-set, the probability $P_{t,k}$ to the probability $P_{n,k}$, and a monotone (increasing or decreasing) property of graphs is identified with a monotone (increasing or decreasing) property of subsets of X.

Note that not every property of the subsets of X corresponds to a graph property, because a graph property is invariant under graph isomorphisms: if $G, H \in \mathcal{G}^t$, $G \in Q$ and $G \cong H$ then $H \in Q$. For example, if $t = 3$, $n = 3$, and we write 1 for the edge x_1x_2, 2 for x_2x_3 and 3 for x_3x_1, and define $Q = \big\{\{1\}, \{1,2\}, \{1,3\}, \{1,2,3\}\big\}$ then Q is a monotone increasing property of subsets of $X = \{1,2,3\}$ but it clearly does not correspond to a graph property on $\mathcal{G}^3$.

Our main aim in this section is to prove a recent theorem of Bollobás and Thomason (1986) stating that not only do many graph properties have threshold functions but, in fact, every non-trivial monotone increasing property of sets has a threshold function. A property Q of subsets of X is *non-trivial* if $Q \ne \emptyset$ and $Q \ne \mathcal{P}(X)$. Thus a monotone increasing property Q is non-trivial iff $\emptyset \notin Q$ and $X \in Q$.

The size of a shadow of a set system is intimately related to the way $P_k(Q) = P_{n,k}(Q)$ is forced to vary with k for a monotone property Q. (To simplify the notation, from now on we write P_k for $P_{n,k}$). Indeed, Q is monotone increasing iff for every $k \le n-1$ the upper shadow $\partial_u Q_k$ of the kth level Q_k is contained in Q_{k+1}, the next level; similarly, Q is monotone decreasing iff $\partial_l Q_k \subset Q_{k-1}$ for every k, $1 \le k \le n$. What does the local LYM inequality (inequality (2) in §3) tell us? If Q is monotone decreasing then

$$|\partial_l Q_r| \Big/ \binom{n}{r-1} \ge |Q_r| \Big/ \binom{n}{r},$$

which is nothing else but

$$P_{r-1}(Q) \ge P_r(Q). \tag{1}$$

Thus the local LYM inequality tells us that for a monotone decreasing property Q the value $P_{n,r}(Q)$ increases as r decreases. What we would like to show is that $P_{n,r}(Q)$ not only increases but increases *fast*. Therefore it is not surprising that we turn to the best possible bound on $|\partial_l Q_r|$ in terms of $|Q_r|$, namely the Kruskal-Katona theorem.

Let $\partial^{(r)}(m)$ be the function defined in (5.8) and appearing in Theorem 5.3. Thus for $m = b^{(r)}(m_r, \ldots, m_s) = \sum_{j=s}^{r} \binom{m_j}{j}$, $m_r > m_{r-1} > \ldots m_s \geq 1$, we have

$$\partial^{(r)}(m) = b^{(r-1)}(m_r, \ldots, m_s) = \sum_{j=s}^{r} \binom{m_j}{j-1}.$$

We shall need a result of Lovász (1979) giving a lower bound for $\partial^{(r)}(m)$ which is easier to use than the exact function itself. This bound makes use of extensions of binomial coefficients. For a natural number r and a real number $x \geq r-1$ define

$$\binom{x}{r} = \frac{x(x-1)\ldots(x-r+1)}{r!}$$

and for $x \geq 0$ let

$$\binom{x}{0} = 1.$$

The polynomial $\binom{x}{r}$ is strictly increasing for $x \geq r-1$, it is 0 at $x = r-1$ and satisfies the relation

$$\binom{x}{r} = \binom{x-1}{r} - \binom{x-1}{r-1}.$$

Theorem 1. *Let $r \geq 1$, $\emptyset \neq \mathcal{A} \subset X^{(r)}$ and define $x \geq r$ by $\binom{x}{r} = |\mathcal{A}|$. Then*

$$|\partial \mathcal{A}| \geq \binom{x}{r-1}.$$

Proof. The assertion can be proved by checking that for $m = \binom{x}{r} \geq 1$ one has

$$\partial^{(r)}(m) \geq \binom{x}{r-1}. \tag{2}$$

In fact, by Theorem 5.3, our assertion is precisely inequality (2). Though checking (2) sounds easy, the manipulations are not too pleasant. It is more elegant to notice that the proof of the Kruskal-Katona theorem can be changed to give (2).

Let then $\mathcal{A}$, $\mathcal{A}_1$ and $\mathcal{A}_0$ be as in the proof of Theorem 5.3 and let us use the same induction hypothesis. In particular, $r \geq 2$. If $|\mathcal{A}_1| < \binom{x-1}{r-1}$ then

$$|\mathcal{A}_0| = |\mathcal{A}| - |\mathcal{A}_1| > \binom{x}{r} - \binom{x-1}{r-1} = \binom{x-1}{r}$$

so

$$|\mathcal{A}_1| \geq |\partial \mathcal{A}_0| > \binom{x-1}{r-1},$$

a contradiction! Therefore $|\mathcal{A}_1| \geq \binom{x-1}{r-1}$ so

$$|\partial \mathcal{A}| \geq |\partial \mathcal{A}_1| \geq |\mathcal{A}_1| + \partial^{(r-1)}(|\mathcal{A}_1|) \geq \binom{x-1}{r-1} + \binom{x-1}{r-2} = \binom{x}{r-1}. \blacksquare$$

As a trivial consequence of Theorem 1, one obtains the following improvement of the local LYM inequality, noted by Bollobás and Thomason (1986).

Theorem 2. *Let $Q \subset \mathcal{P}(X)$ be an ideal. Then for $0 \leq s < r \leq n = |X|$ we have*

$$P_s(Q)^r \geq P_r(Q)^s \tag{3}$$

Proof. If $Q_r = X^{(r)}$ then $Q_s = X^{(s)}$ so both sides of (3) are 1, and if $Q_r = \emptyset$ then (3) is trivial. Assume then that $\emptyset \neq Q \neq X^{(r)}$ and $|Q_r| = \binom{x}{r}$. Since $\frac{x-t}{n-t}$ is a monotone decreasing function of t for $0 \leq t \leq x \leq n$,

$$\left(\frac{x}{n}\right)^{r-s} \left(\frac{x-1}{n-1}\right)^{r-s} \cdots \left(\frac{x-s+1}{n-s+1}\right)^{r-s}$$

$$\geq \left(\frac{x-s}{n-s}\right)^{s} \left(\frac{x-s-1}{n-s-1}\right)^{s} \cdots \left(\frac{x-r+1}{n-r+1}\right)^{s}$$

so

$$\left\{\binom{x}{s} \Big/ \binom{n}{s}\right\}^r \geq \left\{\binom{x}{r} \Big/ \binom{n}{r}\right\}^s.$$

Consequently, by Theorem 1,

$$P_s(Q)^r = P_s(Q_s)^r \geq P_s(\partial^{r-s} Q_r)^r \geq \left\{\binom{x}{s} \Big/ \binom{n}{s}\right\}^r$$

$$\geq \left\{\binom{x}{r} \Big/ \binom{n}{r}\right\}^s = P_r(Q)^s. \qquad \blacksquare$$

Corollary 3. *Let Q be a property of subsets of X and let $k_1 < k < k_2$. If Q is monotone decreasing then*

$$P_{k_2}(Q)^{k/k_2} \leq P_k(Q) \leq P_{k_1}(Q)^{k/k_1}$$

and if Q is monotone increasing then

$$P_{k_1}(Q)^{(n-k)/(n-k_1)} \le P_k(Q) \le P_{k_2}(Q)^{(n-k)/(n-k_2)}.$$

Proof. The first inequality is a reformulation of Theorem 2. To prove the second, note that if Q is monotone increasing then $Q^* = \{A \in \mathcal{P}(X) : X \setminus A \in Q\}$ is monotone decreasing and $P_k(Q) = P_{n-k}(Q^*)$. ∎

The existence of a threshold function is an immediate consequence of Corollary 3; in fact, we not only obtain the existence but an essentially best possible bound for the speed of convergence.

Theorem 4. *Let Q be a monotone increasing non-trivial property of subsets of X, $|X| = n$. Define $k^*(n)$ as*

$$k^*(n) = \max\{l : P_l(Q) \le 1/2\}.$$

Then $k^*(n)$ is a threshold function for Q. Furthermore, if $\omega(n) \ge 1$ and $k \le k^*/\omega$ then

$$P_k(Q) \le 1 - 2^{-1/\omega}$$

and if $k \ge \omega(n)(k^* + 1)$ then

$$P_k(Q) \ge 1 - 2^{-\omega}.$$

Proof. If $k \le k^*/\omega$ then

$$P_k(\neg Q) \ge P_{k^*}(\neg Q)^{1/\omega} \ge 2^{-1/\omega}$$

and if $k \ge \omega(n)(k^* + 1)$ then

$$P_k(\neg Q) \le P_{k^*+1}(\neg Q)^{\omega} \le 2^{-\omega}.$$

These inequalities imply that k^* is a threshold function. ∎

Let us remark on a somewhat different aspect of random sets. If we pick an r-set and an s-set at random then, as a moment's reflection will tell us, the expected size of the intersection is rs/n. In fact, the same assertion is true if we do not choose from all r-sets and all s-sets, not even from collections of r-sets and s-sets, only from two fairly evenly distributed set systems.

Theorem 5. *Let $\mathcal{A} \subset \mathcal{P}(X)$ and $\mathcal{B} \subset \mathcal{P}(X)$ be such that $|\mathcal{A}| = a$, $|\mathcal{B}| = b$, every $x \in X$ is in precisely αa sets belonging to $\mathcal{A}$ and in precisely*

βb sets belonging to $\mathcal{B}$. Then the expected size of the intersection of a random element of $\mathcal{A}$ and a random element of $\mathcal{B}$ is $\alpha\beta n$.

Proof. The expectation is

$$\frac{1}{ab}\sum_{A\in\mathcal{A}}\sum_{B\in\mathcal{B}}|A\cap B| = \frac{1}{ab}\sum\{1 : x\in X, A\ni x, B\ni x, A\in\mathcal{A}, B\in\mathcal{B}\}$$

$$= \frac{1}{ab}\sum_{x\in X}\alpha\beta ab = \alpha\beta n. \qquad \blacksquare$$

This very simple assertion is a result of perhaps the most basic combinatorial argument: averaging. Nevertheless, Theorem 5 and its variants are often very useful. In practice X tends to be not a set of elements but a set system or hypergraph, as in the following example. Given a hypergraph $\mathcal{A}\subset X^{(r)}$, consider all hypergraphs isomorphic to $\mathcal{A}$, with vertex set X. Consider this set as a probability space and call an element of this a random $\mathcal{A}$-system.

Corollary 6. *Let $\mathcal{A},\mathcal{B}\subset X^{(r)}$. The expected number of sets in a random $\mathcal{A}$-system which belong to $\mathcal{B}$ is $|\mathcal{A}||\mathcal{B}|/\binom{n}{r}$.*

Proof. For $\mathcal{C}=\mathcal{A}$ or $\mathcal{B}$, the set of random $\mathcal{C}$-systems is a subset of $\mathcal{P}(X^{(r)}$, and every element of $X^{(r)}$ is in $|\mathcal{C}|/\binom{n}{r}$ random $\mathcal{C}$-systems. The number in question is the expected number of common sets of a random $\mathcal{A}$-system and a random $\mathcal{B}$-system. $\blacksquare$

To conclude this section we turn to a problem of random sets in which the probability in question is small and we wish to show that it is not too small. Let A_1 and A_2 be random elements of $\mathcal{P}(X)$, with all sets taken equiprobable, i.e. with probability 2^{-n}. What is the probability that $A_1\subset A_2$? The sets are picked independently of each other; in particular, $A_1=A_2$ may arise. The probability that $A_2\in X^{(k)}$ is $\binom{n}{k}2^{-n}$ and the probability of choosing a subset A_1 of a fixed set $A_2\in X^{(k)}$ is $2^k2^{-n}=2^{k-n}$. Hence

$$P(A_1\subset A_2)=\sum_{k=0}^{n}\binom{n}{k}2^{k-2n}=2^{-2n}\sum_{k=0}^{n}\binom{n}{k}2^k=(3/4)^n.$$

Suppose now that we take a different probability distribution on $\mathcal{P}(X)$: let the sets in $X^{(\lfloor n/2\rfloor)}$ be equiprobable and let all other sets have probability 0. Then

$$P(A_1\subset A_2)=P\big(A_1=A_2\in X^{(\lfloor n/2\rfloor)}\big)=\binom{n}{\lfloor n/2\rfloor}^{-1}\sim(\pi n/2)^{1/2}\,2^{-n}, \tag{4}$$

which is much smaller than $(3/4)^n$.

How small then can the probability $P(A_1 \subset A_2)$ be if we are allowed to take any probability distribution on $\mathcal{P}(X)$? Baumert, McEliece, Rodemich and Rumsey (1980) proved that this probability cannot be smaller than $\binom{n}{\lfloor n/2 \rfloor}^{-1}$, the probability appearing in (4).

Let us note that the lower bound $\frac{1}{2}\binom{n}{\lfloor n/2 \rfloor}^{-1}$ is trivial. Indeed, if $p_1, p_2, \ldots, p_m$ is a probability distribution on $\{C_1, C_2, \ldots, C_m\}$ then the probability that two elements, picked at random, coincide is

$$\sum_{i=1}^{m} p_i^2 \geq \left\{\sum_{i=1}^{m} p_i\right\}^{-2} \Big/ m = 1/m \tag{5}$$

by the Cauchy-Schwarz inequality. Let then $\mathcal{P}(X) = \bigcup_{i=1}^{m} C_i$ be a partition of $\mathcal{P}(X)$ into $m = \binom{n}{\lfloor n/2 \rfloor}$ chains and let p_i be the probability of C_i, i.e. the probability that our random set is in C_i. By (5) the probability that A_1 and A_2 are picked from the same chain is at least $1/m$. Since $P(A_1 \subset A_2) = P(A_1 \supset A_2)$, by (5) we have

$$P(A_1 \subset A_2 \text{ or } A_1 \supset A_2) = 2P(A_1 \subset A_2) - P(A_1 = A_2)$$

and hence, by(5),

$$\begin{aligned} 2P(A_1 \subset A_2) - P(A_1 = A_2) &= 2P(A_1 \subset A_2 \text{ and } A_1 \neq A_2) \\ &\geq P(A_1 \in C_i \text{ and } A_2 \in C_i \text{ for some } i) \geq 1/m. \end{aligned} \tag{6}$$

Thus

$$P(A_1 \subset A_2) > \tfrac{1}{2} P(A_1 \subset A_2 \text{ or } A_1 \supset A_2) \geq \tfrac{1}{2m}. \tag{7}$$

In order to prove the theorem of Baumert, McEliece, Rodemich and Rumsey (1980), we have to get rid of the factor $\frac{1}{2}$. How can we do this? By considering not only one partition but two orthogonal partitions into chains, i.e. two partitions such that no two sets belong to the same chain in both partitions.

Theorem 7. *Let $n \geq 2$ and let A_1 and A_2 be independent random subsets of X with an arbitrary probability distribution on $\mathcal{P}(X)$. Then*

$$P(A_1 \subset A_2) \geq 1/m$$

where $m = \binom{n}{\lfloor n/2 \rfloor}$.

Proof. Let p_A, $A \in \mathcal{P}(X)$, be the probability distribution on $\mathcal{P}(X)$. By the result of Shearer and Kleitman (1979) in Exercise 4.5, we can find two orthogonal partitions of $\mathcal{P}(X)$ into m chains, say $\mathcal{P}(X) = \bigcup_{i=1}^m C_i = \bigcup_{i=1}^m C_i'$.

Note that

$$P(A_1 \in C_i \text{ and } A_2 \in C_i \text{ for some } i) = \sum_A p_A^2 + 2\sum\nolimits_0 p_A p_B$$

where $\sum_0$ denotes summation over all pairs $A, B \in \mathcal{P}(X)$ such that $A \subset B$, $A \neq B$, and A and B belong to the same chain C_i. Hence, by (6),

$$\sum_A p_A^2 + 2\sum\nolimits_0 p_A p_B \geq 1/m. \tag{8}$$

Also, with the analogous definition of $\sum_0'$ for the second partition,

$$\sum_A p_A^2 + 2\sum\nolimits_0' p_A p_B \geq 1/m. \tag{9}$$

Finally,

$$P(A_1 \subset A_2) = \sum_A p_A^2 + \sum\nolimits^* p_A p_B$$

where $\sum^*$ denotes summation over all pairs $A, B \in \mathcal{P}(X)$ such that $A \subset B$ and $A \neq B$. Since no term $p_A p_B$ appears in both $\sum_0$ and $\sum_0'$, on adding (8) and (9) and dividing by 2, we find that

$$P(A_1 \subset A_2) = \sum_A p_A^2 + \sum\nolimits^* p_A p_B$$

$$\geq \sum_A p_A^2 + \sum\nolimits_0 p_A p_B + \sum\nolimits_0' p_A p_B \geq 1/m. \qquad \blacksquare$$

Theorem 7 is another extension of Sperner's theorem. Indeed, if $\mathcal{F} = \{F_1, F_2, \ldots, F_k\}$ is a Sperner system then take the probability distribution on $\mathcal{P}(X)$ in which each F_i has probability $p_i = 1/k$ and (necessarily) all sets not in $\mathcal{F}$ have probability 0. Then

$$P(A_1 \subset A_2) = P(A_1 = A_2 = F_i \text{ for some } i)$$

$$= \sum_{i=1}^k p_i^2 = 1/k.$$

Hence, by Theorem 7, we have $1/k \geq 1/m$, i.e. $k \leq m$ — precisely Sperner's theorem!

Exercises

1. Check that the probability that a random graph $G_{4,4}$ contains a triangle is $4/5$.

2. Show that the probability that a random graph $G_{5,4}$ is connected is $25/42$. (There are 5^3 trees on $[5]$.)

3. Suppose $k/n^{1/2} \to 0$. Prove that almost no random k-subset of $X = [n]$ contains two consecutive integers.

4. Show that $k^* = n^{1/2}$ is a threshold function for the random set property of containing two consecutive integers.

5. Prove that if $k/n^{1/3} \to 0$ then almost no random k-subset of $[n]$ contains an arithmetic progression of length 3. What is the threshold function of this property?

6. Let Q be the property that a graph of order t contains a triangle. Prove that $k = t$ is a threshold function for Q.

7. Let Q and k^* be as in Theorem 4 and suppose $0 < \underline{\lim} k^*(n)/n \leq \overline{\lim} k^*(n)/n < 1$. Prove that not only is $n/2$ a threshold function for Q but the following stronger assertion holds as well: if $k = o(n)$ then almost no k-set has Q and if $k = n - o(n)$ then almost every k-set has Q.

8. Let $n = 2k + 1$ and define a probability distribution on $\mathcal{P}(X)$ by setting

$$p_A = \begin{cases} \frac{1}{2}\binom{n}{k}^{-1} & \text{if } A \in X^{(k)} \cup X^{(k+1)}, \\ 0 & \text{otherwise.} \end{cases}$$

What is $P(A_1 \subset A_2)$, where A_1 and A_2 are independent random subsets of X?

§7. INTERSECTING HYPERGRAPHS

A set system $\mathcal{A} \subset \mathcal{P}(X)$ is said to be an *intersecting family* if $A_1, A_2 \in \mathcal{A}$ imply $A_1 \cap A_2 \neq \emptyset$. How large can an intersecting family be? It is a little surprising that the answer to this question is entirely trivial: an intersecting family $\mathcal{A}$ of subsets of a set X of size n has at most 2^{n-1} elements since for every subset $A \subset X$ the family contains at most one of the sets A and $X \setminus A$. The set system $\mathcal{P}(X_x) = \{A : A \subset X, x \in A\}$, where $x \in X$ is an intersecting family with 2^{n-1} elements. To get another easily constructed intersecting family of size 2^{n-1}, assume that n is odd and take $\{A : A \subset X, |A| \geq n/2\}$ and add to it precisely one member of each pair $(A, X \setminus A)$, $|A| = |X \setminus A| = n/2$. In fact, every maximal intersecting family has precisely 2^{n-1} elements in it for it is a monotone increasing family and contains precisely one of any two complementary sets.

How large then is an *intersecting hypergraph*? In other words, at most how many edges can an r-uniform hypergraph have if no two of the edges are disjoint? For $n \leq 2r$ this is again a trivial question because $X^{(r)}$ is an intersecting family for $n < 2r$, and if $n = 2r$ then every maximal r-graph contains one of any two complementary sets of size r, so the maximum is $\frac{1}{2}\binom{n}{r} = \binom{n-1}{r-1}$.

What then if $n > 2r$? This case is no longer trivial. Note that for $x \in X$ the r-graph $X_x^{(r)} = \{A \in X^{(r)} : x \in A\}$ is an intersecting family and it is a maximal intersecting r-graph. In a paper which has turned out to be a milestone in the theory of extremal set systems, Erdős, Ko and Rado (1961) proved that $X_x^{(r)}$ is the unique r-graph of maximal size without two disjoint edges.

Theorem 1. *Let $2 \leq r < n/2$ and $\mathcal{A} \subset X^{(r)}$ an intersecting* *hypergraph. Then*

$$|\mathcal{A}| \leq \binom{n-1}{r-1} \tag{1}$$

with equality iff $\mathcal{A} = X_x^{(r)}$ *for some* $x \in X$.

We shall give two proofs, both of them very elegant. The first one, due to Daykin (1974), is based on the Kruskal-Katona theorem, but the second, given by Katona (1972), is self-contained.

1*st Proof.* We shall deduce the result from the Kruskal-Katona theorem, Theorem 5.3. Denote by ∂^t the operation of taking lower shadows t times, i.e. for a hypergraph $\mathcal{F} \subset X^{(s)}$ and $1 \leq t \leq s$ let $\partial^t \mathcal{F} = \partial(\partial(\ldots(\partial \mathcal{F})\ldots)) = \{E \in X^{(s-t)} : E \subset F \text{ for some } F \in \mathcal{F}\}$. Set $\mathcal{B} = \mathcal{A}^c = \{X \setminus A : A \in \mathcal{A}\} \subset X^{(n-r)}$. Then $A \in \mathcal{A}$ and $B \in \mathcal{B}$ imply $A \not\subset B$. Hence $(\partial^{n-2r}\mathcal{B}) \cap \mathcal{A} = \emptyset$.

Let us assume that equality holds in (1). Since $X_x^{(r)}$ is a maximal intersecting family of r-sets, Theorem 1 follows if we show that $\mathcal{A} \simeq X_x^{(r)}$. As $|\mathcal{A}| = |\mathcal{B}| = \binom{n-1}{r-1} = \binom{n-1}{n-r}$, by Theorem 5.3 we have $|\partial\mathcal{B}| \geq \binom{n-1}{n-r-1}$, $|\partial^2\mathcal{B}| \geq \binom{n-1}{n-r-2}, \ldots, |\partial^{n-2r}\mathcal{B}| \geq \binom{n-1}{r}$. Furthermore,

$$|\partial^{n-2r}\mathcal{B}| = \binom{n-1}{r}$$

iff $\mathcal{B} = Y^{(n-r)}$ for some $Y = X \setminus \{x\}$ and $\mathcal{A} = X_x^{(r)}$. ∎

2*nd Proof.* Once again, assume that $|\mathcal{A}| = \binom{n-1}{r-1}$. Arrange the elements of X in a cyclic order. How many of the sets $A \in \mathcal{A}$ can form *intervals*, i.e. sets consisting of consecutive elements? At most r, since if $(a_1, a_2, \ldots, a_r)$ is one of these intervals then for every i, $1 \leq i \leq r-1$, there is at most one interval which separates a_i from a_{i+1}, i.e. contains precisely one of a_i and a_{i+1}.

As there are $\frac{1}{2}(n-1)!$ cyclic orders and every $A \in \mathcal{A}$ is mapped into an interval in $\frac{1}{2}r!(n-r)!$ of these orders, on average there are

$$\tfrac{1}{2}|\mathcal{A}|r!(n-r)!/\left\{\tfrac{1}{2}(n-1)!\right\} = r$$

intervals in a cyclic order. Therefore we must have r intervals in each cyclic order, so in each cyclic order $\mathcal{A}$ must contain sets of the form

$$\{a_1, a_2, \ldots, a_r\},\ \{a_2, a_3, \ldots, a_{r+1}\}, \ldots,\ \{a_r, a_{r+1}, \ldots, a_{2r-1}\} \tag{2}$$

where a_{i+1} follows a_i in the cyclic order.

Suppose then that the sets (2) belong to $\mathcal{A}$. Then for $b \notin \{a_1, a_2, \ldots, a_{2r-1}\} \in \mathcal{A}$, consider a cyclic order of the form $b, a_1, a_2, \ldots, a_r, c_{r+1}, c_{r+2}, \ldots, c_{2r-1}, \ldots$. Since $\{b, a_1, a_2, \ldots, a_{r-1}\} \notin \mathcal{A}$ but $\{a_1, a_2, \ldots, a_r\} \in \mathcal{A}$, all the sets $\{a_2, a_3, \ldots, a_r, c_{r+1}\}$, $\{a_3, a_4, \ldots, a_r, c_{r+1}, c_{r+2}\}, \ldots, \{a_r, c_{r+1}, c_{r+2}, \ldots, c_{2r-1}\}$ must belong to $\mathcal{A}$. But then this easily implies that $\mathcal{A} \supset X_{a_r}^{(r)}$ and so $\mathcal{A} = X_{a_r}^{(r)}$. ■

Note that the second proof can be shortened slightly by quoting Corollary 6.6. Let $J \subset X^{(r)}$ consist of the following n r-sets; $1\,2 \ldots r$, $2\,3 \ldots (r+1), \ldots, (n-1)n \ldots (r-1)$, $n\,1 \ldots (r-1)$. By Corollary 6.6 the expected number of sets of a random J-system which belong to $\mathcal{A}$ is $|J||\mathcal{A}|/\binom{n}{r} = r$. As no J-system can contain more than r-sets of $\mathcal{A}$, all J-systems must contain precisely r sets of $\mathcal{A}$ and we are done as before.

In the theorem above we only demanded that any two sets in our system intersect. What happens if we demand not only that our sets intersect but that the intersection is substantial? Let $1 \leq l < r$. Call a set system $\mathcal{A} \subset \mathcal{P}(X)$ *l-intersecting* if $|A_1 \cap A_2| \geq l$ for all $A_1, A_2 \in \mathcal{A}$. (Note that a 1-intersecting family is precisely an intersecting family.) How large can an l-intersecting family of r-sets be? In other words, at most how many edges can an l-intersecting r-graph have? In studying this question, we may assume that $n > 2r - l$ since if $n \leq 2r - l$ then any two r-sets intersect in at least l elements so $X^{(r)}$ is the (trivial and) unique maximal l-intersecting r-graph. The most obvious way of constructing an l-intersecting r-graph is to imitate $X_x^{(r)}$, namely to take $\mathcal{F}_0 = X^{(r)}(L_0) = \{A \in X^{(r)} : A \supset L_0\}$, where L_0 is some fixed l-subset of X, say $L_0 = [l]$. Let us say that $\mathcal{F} \subset X^{(r)}$ is *fixed by an l-set* if some l-set is contained in all members of $\mathcal{F}$. It is immediate that a system fixed by an l-set is an l-intersecting family, and every maximal family of r-sets fixed by an l-set is of the form $X^{(r)}(L)$ for some $L \in X^{(l)}$. Clearly $|\mathcal{F}_0| = \binom{n-l}{r-l}$ and it is not unreasonable to expect that we cannot get an l-intersecting family of r-sets with more sets. However, $\mathcal{F}_0$ is not the only natural candidate for a large l-intersecting r-graph. For $0 \leq t \leq r - l$, let

$$\mathcal{F}_t = \{A \in X^{(r)} : |A \cap L_t| \geq l + t\}$$

where L_t is some fixed $(l+2t)$-subset of X, say $L_t = [l + 2t]$. (Strictly speaking, up to isomorphism, $\mathcal{F}_t$ depends on n, r, l and t, but we suppress the dependence on n, r and l.) Then for $A_1, A_2 \in \mathcal{F}_t$ we have $|A_1 \cap A_2| \geq |(A_1 \cap L_t) \cap (A_2 \cap L_t)| \geq (l+t) + (l+t) - (l+2t) = l$ so $\mathcal{F}_t$ is an l-intersecting r-graph. Clearly

$$|\mathcal{F}_t| = \sum_{s=l+t}^{r} \binom{l+2t}{s} \binom{n-l-2t}{r-s}.$$

Which of the l-intersecting r-graphs $\mathcal{F}_0, \mathcal{F}_1, \ldots, \mathcal{F}_{r-l}$ have the greatest number of elements? The answer depends on the size of n. For example, if $r = 5$ and $l = 3$ then for $n = 8$ we have $|\mathcal{F}_0| = 10$, $|\mathcal{F}_1| = 21$, so $|\mathcal{F}_2|$ is the largest, for $n = 11$ the sizes are 23, 31, 21, so $|\mathcal{F}_1|$ is the largest and for $n = 13$ we have $|\mathcal{F}_0| = 45$, $|\mathcal{F}|_1 = 41$ and $|\mathcal{F}_2| = 21$, so $|\mathcal{F}_0|$ is the largest. In fact, $|\mathcal{F}_{r-l}|$ is independent of n and as n grows, eventually $|\mathcal{F}_0|$ becomes the largest. Erdős, Ko and Rado (1961) proved that if n is sufficiently large then, up to isomorphism, $\mathcal{F}_0$ is the unique l-intersecting r-graph of maximal size. We shall deduce this from the following bound on the size of an l-intersecting r-graph not fixed by an l-set.

Theorem 2. *Suppose $2 \leq l < r$, $n > 2r - l$ and $\mathcal{A} \subset X^{(r)}$ is an l-intersecting r-graph not fixed by an l-set. Then*

$$|\mathcal{A}| \leq r\binom{n-l-1}{r-l-1} + \sum_{t=1}^{l}\binom{l}{t}\binom{r-l}{t}^2\binom{n-l-t}{r-l-t}. \tag{3}$$

Proof. We may and shall assume that $\mathcal{A}$ is a maximal l-intersecting r-graph. Then there are edges $A_1, A_2 \in \mathcal{A}$ such that $|A_1 \cap A_2| = l$. Since $B = A_1 \cap A_2$ does not fix $\mathcal{A}$, there is an edge $A_3 \in \mathcal{A}$ with $A_1 \cap A_2 \cap A_3 \neq B$. Let

$$\mathcal{A}_t = \{A \in \mathcal{A} : |B \setminus A| = t\}, t = 0, 1, \ldots, l.$$

How large can a set system $\mathcal{A}_t$ be? If $t \geq 1$ then $|A \cap (A_1 \setminus B)| \geq t$ and $|A \cap (A_2 \setminus B)| \geq t$ for every $A \in \mathcal{A}_t$ so, rather crudely,

$$|\mathcal{A}_t| \leq \binom{l}{t}\binom{r-l}{t}^2\binom{n-l-t}{r-l-t} = S_t. \tag{4}$$

If $A \in \mathcal{A}_0$ then $|A \cap (A_3 \setminus B)| \geq 1$ so

$$|\mathcal{A}_0| \leq r\binom{n-l-1}{r-l-1}. \tag{5}$$

Since $\mathcal{A} = \bigcup_{t=0}^{l} \mathcal{A}_t$, inequalities (4) and (5) imply (3). ∎

From Theorem 2 it is easy to deduce that if n is sufficiently large then $\mathcal{F}_0$ is the unique l-intersecting r-graph of maximal cardinality.

Theorem 3. *If $2 \le l < r$, $n \ge 2lr^3$ and $\mathcal{A} \subset X^{(r)}$ is an l-intersecting hypergraph then*

$$|\mathcal{A}| \le \binom{n-l}{r-l},$$

with equality iff $\mathcal{A} \simeq \mathcal{F}_0 = \{A \in X^{(r)} : A \supset [l]\}$.

Proof. If $\mathcal{A}$ is fixed by an l-set L then every $A \in \mathcal{A}$ is determined by $A \setminus L \in (X \setminus L)^{(r-l)}$ so $|\mathcal{A}| \le \binom{n-l}{r-l}$, with equality iff $\mathcal{A} \simeq \mathcal{F}_0$.

On the other hand, if $\mathcal{A}$ is not fixed by an l-set then, by Theorem 2, inequality (3) holds. Hence all we need is that the right-hand-side of (3) is less than $\binom{n-l}{r-l}$, i.e.

$$\begin{aligned}\sum_{t=2}^{l} S_t &= \sum_{t=2}^{l} \binom{l}{t}\binom{r-l}{t}^2\binom{n-l-t}{r-l-t} \\ &< \binom{n-l}{r-l} - \{r-l+l(r-l)^2\}\binom{n-l-1}{r-l-1} \\ &= \binom{n-l-1}{r-l-1}\left\{\frac{n-l}{r-l} - r - l(r-l)^2\right\}. \qquad (6)\end{aligned}$$

Since for $2 \le t < l$ we have

$$S_{t+1}/S_t = \frac{l-t}{t+1}\left(\frac{r-l-t}{t+1}\right)^2 \frac{r-l-t}{n-l-t} < (t+1)^{-3},$$

inequality (6) is easily seen to hold:

$$\begin{aligned}\sum_{t=2}^{l} S_t &< l^2(r-l)^4\binom{n-l-2}{r-l-2} < \frac{l^2(r-l)^5}{n-l}\binom{n-l-1}{r-l-1} \\ &< \frac{1}{4}\frac{n-l}{r-l}\binom{n-l-1}{r-l-1} < \binom{n-l-1}{r-l-1}\left\{\frac{n-l}{r-l} - r - l(r-l)^2\right\}.\end{aligned}$$

∎

The bound $2lr^3$ in Theorem 3 is somewhat better than the original bound given by Erdős, Ko and Rado, but it is far from being best possible. In view of our cavalier estimates used in the proof of Theorem 2, this is hardly surprising. Frankl (1978) conjectured that one of the l-intersecting r-graphs $\mathcal{F}_0, \mathcal{F}_1, \ldots, \mathcal{F}_{r-l}$ has maximal cardinality among all l-intersecting r-graphs, i.e. if $\mathcal{A} \subset X^{(r)}$ is l-intersecting then

$$|\mathcal{A}| \le \max_{0 \le t \le r-l} |\mathcal{F}_t|. \qquad (7)$$

As we saw before Theorem 2, it depends on n, which of the r-graphs $\mathcal{F}_0, \mathcal{F}_1, \ldots, \mathcal{F}_t$ has maximal cardinality. It is easily checked that $|\mathcal{F}_0| = \max_{0 \le t \le r-l} |\mathcal{F}_t|$ iff $n \ge (l+1)(r-l+1)$ (see Ex. 3) so the most important part of Frankl's conjecture is that if $n \ge (l+1)(r-l+1)$ then

$$|\mathcal{A}| \le \binom{n-l}{r-l}$$

for every l-intersecting r-graph $\mathcal{A}$. This deep assertion was proved by Frankl (1978) for $r \ge 15$. Furthermore, Frankl showed that if $r \ge 15$ and $n > (l+1)(r-l+1)$ then $\mathcal{F}_0$ is the only extremal hypergraph and if $r \ge 15$ and $n = (l+1)(r-l+1)$ then there are precisely two extremal hypergraphs: $\mathcal{F}_0$ and $\mathcal{F}_1$. For a wealth of material related to the Erdős-Ko-Rado theorem, the reader should consult Deza and Frankl (1983).

A pleasing application of the first Erdős-Ko-Rado theorem is a result of Liggett (1977) concerning sums of random variables. A random variable (r.v.) Z is a *Bernoulli r.v. with mean p* if Z takes only two values, 0 and 1, and $P(Z=1) = p$, $P(Z=0) = 1-p$. Thus a Bernoulli r.v. can be thought of as the outcome of tossing a biased coin.

Theorem 4. *Let $Z_1, Z_2, \ldots, Z_n$ be independent Bernoulli r.vs each having mean $p \ge 1/2$, and let $c_1, c_2, \ldots, c_n$ be positive numbers such that $\sum_{i=1}^n c_i = 1$. Then*

$$P(\sum_{i=1}^n c_i Z_i \ge 1/2) \ge p$$

Proof. Let $\mathcal{F} = \{A \subset X : \sum_{i \in A} c_i \ge 1/2\}$, $\mathcal{F}_k = \mathcal{F} \cap X^{(k)}$ and $f_k = |\mathcal{F}_k|$. Then, as our r.vs are independent,

$$P(\sum_{i=1}^n c_i Z_i \ge 1/2) = \sum_{k=0}^n f_k p^k (1-p)^{n-k}. \tag{8}$$

What do we know about the f_k's? Firstly, for $A \subset X$ at least one of A and $X \setminus A$ belongs to $\mathcal{F}$ so

$$f_k + f_{n-k} \ge \binom{n}{k}. \tag{9}$$

In particular, if $k = n/2$ (and so n is even) then

$$f_k \ge \frac{1}{2}\binom{n}{k} = \binom{n-1}{k-1}. \tag{10}$$

Secondly, for $k < n/2$ the hypergraph $\mathcal{F}_k$ is intersecting since if $A, B \in \mathcal{F}_k$ then $A \cap B = \emptyset$ implies that $A \cup B \neq X$ so

$$1 > 1 - \sum_{i \notin A \cup B} c_i = 1 - \sum_{i \in A} c_i - \sum_{i \in B} c_i \geq 1/2 + 1/2,$$

a contradiction! Hence, by Theorem 1, for $k < n/2$ we have

$$f_k \leq \binom{n-1}{k-1}. \tag{11}$$

The proof is almost complete. As $p \geq 1/2$ for $k \leq n/2$ the coefficient of f_k in (8) is at most as large as the coefficient of f_{n-k}:

$$p^k(1-p)^{n-k} \leq p^{n-k}(1-p)^k. \tag{12}$$

In order to treat the odd n and even n cases together, for n odd set $f_{n/2} = 0$. Then, by (9), (10) and (11),

$$P(\sum_{i=1}^{n} c_i Z_i \geq 1/2) = \sum_{k<n/2} (f_k + f_{n-k}) p^{n-k}(1-p)^k +$$

$$\sum_{k<n/2} f_k \{p^k(1-p)^{n-k} - p^{n-k}(1-p)^k\} + f_{n/2} p^{n/2}(1-p)^{n/2}$$

$$\geq \sum_{k<n/2} \binom{n}{k} p^{n-k}(1-p)^k$$

$$+ \sum_{k<n/2} \binom{n-1}{k-1} \{p^k(1-p)^{n-k} - p^{n-k}(1-p)^k\} + f_{n/2} p^{n/2}(1-p)^{n/2}$$

$$\geq \sum_{k=1}^{n} \binom{n-1}{k-1} p^k (1-p)^{n-k}$$

$$= p \sum_{k=1}^{n} \binom{n-1}{k-1} p^{k-1}(1-p)^{n-1-(k-1)} = p.$$

■

Exercises

1. Let $\mathcal{F} \subset \mathcal{P}(X)$ be a maximal family such that if $A, B \in \mathcal{F}$ then $A \cup B \neq X$. Prove that $\mathcal{F}$ is a monotone decreasing family (an ideal) and $|\mathcal{F}| = 2^{n-1}$.

2. Prove the following extension of Theorem 1. If $1 \leq r < n/2$ and $\mathcal{F} \subset \bigcup_{j=0}^{r} X^{(j)}$ is an intersecting family of sets of size at most r then

$$|\mathcal{A}| \leq \binom{n-1}{r-1}.$$

3. Let $\mathcal{F}_0, \mathcal{F}_1, \ldots, \mathcal{F}_{r-l}$ be the set systems defined before Theorem 2. Prove that

$$|\mathcal{F}_0| > \max_{1 \leq t \leq r-l} |\mathcal{F}_t| \quad \text{if } n > (l+1)(r-l+1),$$

$$|\mathcal{F}_0| = |\mathcal{F}_1| \quad \text{if } n = (l+1)(r-l+1)$$

and

$$|\mathcal{F}_0| < |\mathcal{F}_1| \quad \text{if } n < (l+1)(r-l+1).$$

4. Let $1 \leq s < r < n$ and let $\mathcal{F} \subset X^{(r)}$ be a hypergraph such that $|A_1 \cap A_2| \leq s$ whenever $A_1, A_2 \in \mathcal{F}$ and $A_1 \neq A_2$. Show that

$$|\mathcal{F}| \leq (n)_{s+1}/(r)_{s+1}.$$

5. Let $r = p+1$, $n = p^2 + p + 1$ and let $\mathcal{F} \subset X^{(r)}$ be such that $|A_1 \cap A_2| \leq 1$ whenever $A_1, A_2 \in \mathcal{F}$ and $A_1 \neq A_2$. Prove that $|\mathcal{F}| \leq n$ and equality holds iff $\mathcal{F}$ is the set of lines of a projective plane of order p, i.e. a projective plane on $n = p^2 + p + 1$ points with $r = p + 1$ points on a line. Show also that for $p = 2$ equality holds for a unique hypergraph $\mathcal{F}$ (up to isomorphism, as always).

§8. THE TURÁN PROBLEM

Many of the problems we have studied so far have the following form. Given a property Q and an invariant μ for a class $\mathcal{H}$ of set systems, what is the maximum (minimum) of μ over all set systems in $\mathcal{H}$ satisfying Q? For example, what is the maximum of $|\mathcal{F}|$ if $\mathcal{F} \subset \mathcal{P}(X)$ and no member of $\mathcal{F}$ is contained in another member of $\mathcal{F}$? For a set system $\mathcal{F}$, let $\mu(\mathcal{F})$ be the number of $(r-1)$-sets contained in at least one member of $\mathcal{F}$. What is the minimum of $\mu(\mathcal{F})$ as $\mathcal{F}$ runs over all set systems $\mathcal{F} \subset X^{(r)}$ with m members? How large can $|\mathcal{F}|$ be if $\mathcal{F} \subset \mathcal{P}(X)$ and any two members of $\mathcal{F}$ intersect? What is the corresponding maximum if $\mathcal{F} \subset X^{(r)}$? All these questions are examples of so-called *extremal problems*. The set systems (hypergraphs) on which the extremum is attained are the *extremal set systems (extremal hypergraphs)* of the problem. We say that there is a *unique* extremal set system (hypergraph) if all extremal set systems (hypergraphs) are isomorphic.

As far as the developments of the subjects are concerned, the theory of hypergraphs (and set systems, for that matter) is a younger brother of graph theory. A central and well-developed part of graph theory is the theory of extremal problems (see Bollobás (1978) for a detailed account of the subject). Therefore it is natural to try to tackle the hypergraph problems corresponding to the fundamental problems in extremal graph theory. The prime example of an extremal problem in graph theory is the *forbidden subgraph problem*: given a graph F, determine $ex(n; F)$, the maximal number of edges in a graph of order n which does not contain F as a subgraph. As customary, we say that G *contains* F if G contains a subgraph isomorphic to F; the same convention is used for hypergraphs and set systems. The graph F is the *forbidden subgraph.*

To be more specific, the starting point of extremal graph theory is the following theorem of Turán (1941) (see also Bollobás (1978, p. 294)) concerning the case when F is K^s, the complete graph of order $s \geq 3$:

$$ex(n;\, K^s) = \binom{n}{2} - \sum_{i=0}^{s-2} \binom{\lfloor (n+i)/(s-1) \rfloor}{2}$$

and the unique extremal graph is $T_{s-1}(n)$, the complete $(s-1)$-partite graph whose classes are as equal as possible. Thus $T_{s-1}(n) = (V, E)$ has the following form: $V = \{1, 2, \dots, n\} = \bigcup_{i=0}^{s-2} V_i$, $|V_i| = \lfloor (n+i)/(s-1) \rfloor$ and E consists of all the edges joining different classes V_i and V_j.

The case when F is a triangle is particularly pleasant: $ex(n;\, K^3) = \lfloor n^2/4 \rfloor$ and the only extremal graph is $T_2(n) = K(\lfloor n/2 \rfloor, \lceil n/2 \rceil)$, the complete bipartite graph with $\lfloor n/2 \rfloor$ vertices in one class and $\lceil n/2 \rceil$ in the other.

In view of this, the following *problem of Turán* can be considered to be the basic extremal hypergraph problem: given $3 \leq r < s \leq n$, what is $ex(n;\, K_s^{(r)})$, the maximal number of edges in an r-graph of order n not containing a $K_s^{(r)}$? Here $K_s^{(r)}$ is a *complete r-graph of order s*: $K_s^{(r)} = (W, W^{(r)})$ where $|W| = r$. As the total number of possible edges is $\binom{n}{r}$, for large values of n one tends to study $\delta(n;\, K_s^{(r)}) = ex(n;\, K_s^{(r)})/\binom{n}{r}$, the proportion of the possible edges an extremal hypergraph contains, i.e. the *density* of the extremal hypergraph. The intricacy of hypergraphs is strongly demonstrated by the fact that the Turán problem for graphs has an easy and elegant solution for all s and n, but for no r and s, $3 \leq r < s$, has the function $ex(n;\, K_s^{(r)})$ been determined for infinitely many values of n. Even more, though it is easily shown that $\lim_{n\to\infty} \delta(n;\, K_s^{(r)}) = \gamma(r, s)$ exists for all values of r and s, for no pair (r, s), $3 \leq r < s$, is the value of $\gamma(r, s)$ known.

We shall discuss briefly some of the simplest of the numerous partial results concerning the function $ex(n;\, K_s^{(r)})$ and then we shall prove a result closely related to the Turán problem. First let us state a natural extension of the Turán problem. Given fixed r-graphs $F_1, F_2, \dots, F_l$, how large is $ex(n;\, F_1, F_2, \dots, F_l)$, the maximal number of edges in an r-graph not containing any of the graphs $F_1, F_2, \dots, F_l$? Here, $F_1, F_2, \dots, F_l$ are the *forbidden subgraphs*. The following results were noted by Katona, Nemetz and Simonovits (1964).

Recall that $(x)_t = x(x-1)\dots(x-t+1)$ is the falling factorial and the size of a hypergraph is the number of edges. One of the most powerful (and very simple) combinatorial arguments, namely averaging, gives the following result.

Theorem 1. *Let $2 \leq r \leq n_0 < n$ and let G be an r-graph of order*

n and size m. Then G contains an r-graph of order n_0 and size at least $m\,(n_0)_r/(n)_r$.

Proof. Let $X_1, X_2, \ldots, X_u$, $u = \binom{n}{n_0}$, be the n_0-subsets of the vertex set X of G, and let m_i be the number of edges contained in X_i. Every r-subset of X and so every edge of G is contained in $v = \binom{n-r}{n_0-r}$ sets X_i. Therefore

$$\sum_{i=1}^{u} m_i = vm$$

and so

$$\max_i m_i \geq vm/u = m\,(n_0)_r/(n)_r. \qquad \blacksquare$$

Theorem 2. *Let $2 \leq r \leq n_0 < n$ and $ex(n_0; F_1, \ldots, F_l) \leq m_0$. Then*

$$ex(n;\, F_1, \ldots, F_l) \leq \lfloor m_0(n)_r/(n_0)_r \rfloor = \left\lfloor \left\{ m_0 \Big/ \binom{n_0}{r} \right\} \binom{n}{r} \right\rfloor.$$

Proof. Let $m = ex(n;\, F_1, \ldots, F_l)$ and consider an r-graph G of order n and size m not containing any of the F_i's. By Theorem 1 the hypergraph G has a subgraph H of order n_0 and size at least $m(n_0)_r/(n)_r$. Since H does not contain any of the F_i's,

$$m(n_0)_r/(n)_r \leq ex(n_0;\, F_1, \ldots, F_l) \leq m_0$$

so

$$m \leq \lfloor m_0(n)_r/(n_0)_r \rfloor. \qquad \blacksquare$$

This result is, once again, a special case of Corollary 6.6: the expected number of edges in a random n_0-set is $m\binom{n_0}{r} \Big/ \binom{n}{r} = m(n_0)_r(n)_r$. Hence some n_0-set contains at least $m(n_0)_r/(n)_r$ edges of G.

Clearly, Theorem 2 carries over to infinite families of forbidden subgraphs (see Ex. 1). Furthermore, there is no reason why we should not make use of bounds on the number of edges in arbitrary subgraphs rather than restrict our attention to complete subgraphs. Here is a rather general though very simple result in this spirit. A collection Q of set systems on X is said to be a *property of set systems* if it is invariant under isomorphism.

Theorem 3. *Let $\mathcal{A} = \bigcup_{i=0}^{n} \mathcal{A}_i$, $\mathcal{A}_i \subset X^{(i)}$, $a_i = |\mathcal{A}_i|$, and let $u_0, u_1, \ldots, u_n$ be non-negative reals. Set $w_i = a_i u_i \Big/ \binom{n}{i}$, $0 \leq i \leq n$.*

Furthermore, for $\mathcal{F} = \bigcup_{i=0}^{n} \mathcal{F}_i$, $\mathcal{F}_i \subset X^{(i)}$, $f_i = |\mathcal{F}_i|$, *define* $u(\mathcal{F}) = \sum_0^n f_i u_i$ *and* $w(\mathcal{F}) = \sum_0^n f_i w_i$.

Let Q *be a property of set systems such that if* $\mathcal{A}' \subset \mathcal{A}$ *has* Q *then* $u(\mathcal{A}') \le 1$. *Then if* $\mathcal{F} \subset \mathcal{P}(X)$ *has* Q *then* $w(\mathcal{F}) \le 1$.

Proof. By Corollary 6.6 (or from first principles) the expected u-weight of a random $\mathcal{F}$-system in $\mathcal{A}$ is

$$\sum_0^n u_i a_i f_i \bigg/ \binom{n}{i} = \sum_0^n f_i w_i = w(\mathcal{F}).$$

But the expectation is at most the maximum, which is 1. ∎

For a property Q of set systems $f(n, Q)$ is the maximal number of elements in a set system on X which has Q.

Corollary 4. *Let* Q *be a monotone decreasing property of* r*-graphs. Suppose there is an* r*-graph* $\mathcal{A} \subset X^{(r)}$ *such that if* $\mathcal{A}' \subset \mathcal{A}$ *has* Q *then* $|\mathcal{A}'| \le \delta|\mathcal{A}|$. *Then* $f(n, Q) \le \delta\binom{n}{r}$. ∎

Note that we have already applied special cases of Theorem 3 and Corollary 4. To prove the LYM inequality (Theorem 3.2), we took $u_i = 1$ and $\mathcal{A} = \{\emptyset, 1, 12, \ldots, 12\ldots n\}$ in Theorem 3, and in the second proof of the Erdős-Ko-Rado theorem (Theorem 7.1) we took $\mathcal{A} = \{12\ldots r, 23\ldots(r+1), \ldots, (n-1)1\ldots(r-1)\}$. But let us return to the function $ex(n; K_s^{(r)})$.

Theorem 5. $\gamma(r,s) = \lim_{n\to\infty} \delta(n; K_s^{(r)}) = \lim_{n\to\infty} ex(n; K_s^{(r)})/\binom{n}{r}$ *exists for all* $2 \le r < s$.

Proof. Theorem 2 tells us precisely that $\delta(n; K_s^{(r)})$ is a monotone decreasing function of n. Hence

$$\underline{\lim}_{n\to\infty} \delta(n; K_s^{(r)}) = \lim_{n\to\infty} \delta(n; K_s^{(r)}) = \gamma(r,s).$$ ∎

Theorem 2 can be used to give an upper bound for $\gamma(r,s)$; we have also the following simple lower bound.

Theorem 6. *For* $n_0 \ge s > r$ *we have*

$$\gamma(r,s) \ge r!\, ex(n_0; K_s^{(r)})/n_0^r.$$

Proof. Let G_0 be an extremal graph of order n_0. For $p \geq 1$ construct an r-graph as follows: replace each vertex of G_0 by p vertices and for each edge of G_0 take all p^r edges having one vertex in each of the r classes corresponding to the vertices of the edge. Clearly G does not contain a $K_s^{(r)}$ and has p^r times as many edges as G_0. Hence

$$\gamma(r,s) \geq \lim_{n\to\infty} ex(n_0; K_s^{(r)})p^r \Big/ \binom{pn_0}{r} = r!\, ex(n_0; K_s^{(r)})/n_0^r. \qquad \blacksquare$$

By Turán's theorem for graphs, $\gamma(2,s) = \frac{s-2}{s-1}$ for $s \geq 3$ but as we remarked earlier, $\gamma(r,s)$ is not known for any pair (r,s), $3 \leq r < s$. Nevertheless, using Theorem 2 and various constructions, in some cases one may obtain fairly good bounds on $\gamma(r,s)$.

Let us have another look at the very simplest case of Turán's theorem, the case of a triangle K^3. Note that a triangle is formed by three edges such that the symmetric difference of two is contained in the third. This prompted Katona to propose the following generalization of the triangle case of Turán's problem: determine the maximal number of edges in an r-graph of order n such that

($\triangle$) the symmetric difference of two edges is not contained in a third.

The main aim of this section is to present a theorem of Bollobás (1974) solving this problem for $r = 3$. As we shall see, the extremal 3-graphs are the exact analogues of the complete bipartite graphs, the extremal graphs for triangles.

Consider a complete 3-partite 3-graph of order n: $X = X_1 \cup X_2 \cup X_3$ is a partition of $X = [n]$ and the hypergraph consists of all the triples meeting each X_i in one vertex. This graph has $|X_1|\,|X_2|\,|X_3|$ edges and satisfies ($\triangle$). Now $|X_1|\,|X_2|\,|X_3|$ is maximized if $|X_1|$, $|X_2|$ and $|X_3|$ are as equal as possible, say $|X_1| = \lfloor n/3 \rfloor$, $|X_2| = \lfloor (n+1)/3 \rfloor$ and $|X_3| = \lfloor (n+2)/3 \rfloor$. Denote this graph by G_0.

Theorem 7. *Let $G = (X, \mathcal{E})$ be a 3-graph satisfying ($\triangle$). Then $|\mathcal{E}| \leq \lfloor n/3 \rfloor \lfloor (n+1)/3 \rfloor \lfloor (n+2)/3 \rfloor$.*

Proof. Let G be a 3-graph satisfying condition ($\triangle$) with $|\mathcal{E}| = \lfloor n/3 \rfloor \lfloor (n+1)/3 \rfloor \lfloor (n+2)/3 \rfloor$. Since G_0 is a maximal graph satisfying ($\triangle$), it suffices to prove that $G \cong G_0$.

For $x, y \in X$, $x \neq y$, set

$$A(x,y) = \{z : (x,y,z) \in \mathcal{E}\},$$

i.e. let $A(x,y)$ be the set of 'copies' of triples in $\mathcal{E}$ with 'base' (x,y). Let $x_0, y_0 \in X$ be such that $A = A(x_0, y_0)$ is maximal. Let B be a set of maximal cardinality among the sets $A(x,y)$, $x \in A$, and let C be a set of maximal cardinality among the sets $A(x,y)$, $x \in A$, $y \in B$. Set $D = X \setminus (A \cup B \cup C)$.

Property $(\triangle)$ means that no triple in $\mathcal{E}$ meets a set $A(x,y)$ in more than one element. In particular,

$(\triangle')$ A, B and C are disjoint and no triple in $\mathcal{E}$ contains two vertices of any of the sets A, B and C.

Set $a = |A|$, $b = |B|$ and $c = |C|$. Then $a \geq b \geq c \geq 0$ and $a + b + c \leq n$. By $(\triangle')$ we have $|\mathcal{E} \cap (A \cup B \cup C)^{(3)}| \leq abc$. Hence it suffices to show that $a+b+c = n$ since $abc \geq \lfloor n/3 \rfloor \lfloor (n+1)/3 \rfloor \lfloor (n+2)/3 \rfloor$ and $a + b + c \leq n$ and $a \geq b \geq c$ imply $a = \lfloor (n+2)/3 \rfloor$, $b = \lfloor (n+1)/3 \rfloor$, $c = \lfloor n/3 \rfloor$.

By $(\triangle')$, there are at most

$$\binom{n-a}{2} - \binom{b}{2} - \binom{c}{2} \tag{1}$$

pairs in $(X \setminus A)^{(2)}$ contained in some triple $\mathcal{E}$. Also

$$|A(x,y)| \leq c \qquad \text{for } x \in A,\ y \in B \tag{2}$$

$$|A(x,y)| \leq b \qquad \text{for } x \in A,\ y \in C \cup D \tag{3}$$

$$|A(x,y)| \leq a \qquad \text{for } x, y \in X \setminus A. \tag{4}$$

Hence, counting each triple in $\mathcal{E}$ three times, once for each pair it contains, we find that

$$3\lfloor n/3 \rfloor \lfloor (n+1)/3 \rfloor \lfloor (n+2)/3 \rfloor = 3|\mathcal{E}|$$

$$= \sum_{x \in A, y \in B} |A(x,y)| + \sum_{x \in A, y \in C \cup D} |A(x,y)| + \sum_{(x,y) \in (X \setminus A)^{(2)}} |A(x,y)|$$

$$\leq abc + a(n-a-b)b + \left\{ \binom{n-a}{2} - \binom{b}{2} - \binom{c}{2} \right\} a. \tag{5}$$

All we have to do then is to show that (5) implies $a + b + c = n$. In order to avoid inessential complications, let us assume that $n = 3k$. Then, multiplying (5) by 2, we find that

$$\begin{aligned} 6k^3 &\le a\{9k^2 + b(6k - 2a - 3b + 2c + 1) - (6k - 1)a + a^2 - 3k - c^2 + c\} \\ &= af(a,b,c), \end{aligned} \tag{6}$$

say. Note that for $a = b = c = k$ we have equality in (6).

Suppose that, contrary to the desired relation, $a + b + c < n$. The right-hand-side of (6) increases strictly with c for $c \le b$. As we increase c, keeping $c \le b$, we cannot arrive at $a+b+c = n$ because for $a+b+c = n$ equality holds in (6). Hence $a + 2b < n$ and (6) holds for $c = b$:

$$\begin{aligned} 6k^3 &\le a\{9k^2 + b(6k - 2a - 2b + 2) - (6k - 1)a + a^2 - 3k\} \\ &= af(a,b,b) = ag(a,b), \end{aligned} \tag{7}$$

say. Now let's take the partial derivative of g with respect to b: $\frac{\partial g}{\partial b} = 6k - 2a + 2 - 4b > 0$. Therefore strict inequality must hold in (7) if we increase b to $b_0 = \frac{3k-a}{2}$:

$$\begin{aligned} 6k^3 &< ag(a,b_0) = af(a,b_0,b_0) \\ &= \frac{3a}{2}(3k - a)^2. \end{aligned}$$

However, this is impossible since the maximum of the right-hand-side is $6k^3$. ∎

Exercises

1. Let $F_1, F_2, \ldots$ be r-graphs. Let us say that an r-graph has property Q if it contains no F_i as a subgraph. Prove that if $2 \le r \le n_0 < n$ and $f(n_0, Q) \le m_0$ then

$$f(n,Q) \le \lfloor m_0 (n)_r / (n_0)_r \rfloor.$$

2. By imitating the proof of Theorem 5 show that $\gamma(Q) = \lim_{n\to\infty} f(n,Q)/\binom{n}{r}$ exists for every monotone decreasing property Q of r-graphs.

3. Show that $ex(5;\, K_4^{(3)}) = 7$ and determine the unique extremal hypergraph. Deduce that $ex(n;\, K_4^{(3)}) \le 7n(n-1)(n-2)/60$ for every $n \ge 5$.

3. Deduce from Ex. 4 that $ex(6; K_4^{(3)}) = 14$ and the unique extremal hypergraph is isomorphic to the 3-graph on [6] containing all edges but 123, 124, 345, 346, 561 and 562.

5. Let $X = X_1 \cup X_2 \cup X_3$ be a partition and let G be the 3-graph on X containing all triples but those having two elements in X_i and one in X_{i+1}, $i = 1, 2, 3$, where $X_4 = X_1$. Check that G does not contain a $K_4^{(3)}$. Deduce that for $n = 3k$ we have

$$ex(n; K_4^{(3)}) \geq \frac{n^2}{54}(5n - 9).$$

Deduce also that

$$5/9 \leq \gamma(3, 4) \leq 7/10.$$

6. Show that $ex(7; K_4^{(3)}) = 23$, $ex(8; K_4^{(3)}) = 36$ and $ex(9; K_4^{(3)}) = 54$ and so improve the upper bounds in the previous exercise to $ex(n; K_4^{(3)}) \leq \frac{9}{14}\binom{n}{3}$ if $n \geq 9$ and $\gamma(3, 4) \leq 9/14$. (Katona, Nemetz and Simonovits (1964))

7. Following de Caen (1983), give yet another improvement on the upper bound for $ex(n; K_4^{(3)})$. Let $\mathcal{F} \subset X^{(3)}$ be $K_4^{(3)}$-free with $m = ex(n; K_4^{(3)}) = |\mathcal{F}|$ and let $A(x, y)$ be as in the proof of Theorem 7. Note that if $xyz \in \mathcal{F}$ then $A(x, y) \cap A(y, z) \cap A(z, x) = \emptyset$ and so

$$a_{xy} + a_{yz} + a_{zx} \leq 2n - 3$$

where $a_{uv} = |A(u, v)|$. Summing over all edges in $\mathcal{F}$, deduce that

$$\sum_{xy \in V^{(2)}} d_{xy}^2 \leq (2n - 3)m.$$

Use Cauchy's inequality to show that the left-hand-side is at least

$$\Big(\sum_{xy \in V^{(2)}} d_{xy}\Big)^2 \Big/ \binom{n}{2} = (3m)^2 \Big/ \binom{n}{2}$$

and deduce that $m = ex(n; K_4^{(3)}) \leq \frac{2n-3}{9}\binom{n}{2}$ and $\gamma(3, 4) \leq 2/3$.

8. Let $X = X_1 \cup X_2$ be a partition with $|X_1| = \lfloor n/2 \rfloor$ and $|X_2| = \lceil n/2 \rceil$ and let G be the 3-graph on X containing all triples but those contained in X_1 or X_2. Check that G does not contain a $K_5^{(3)}$ and deduce that

$$\gamma(3, 5) \geq 3/4.$$

What upper bounds can you give for $\gamma(3, 5)$?

§9. SATURATED HYPERGRAPHS

An r-graph $G = (X, \mathcal{E})$ is said to be $(r+s)$-*saturated* if $\mathcal{E} \not\supset Y^{(r)}$ for every $Y \subset X$, $|Y| = r+s$ but for every $F \in X^{(r)} \setminus \mathcal{E}$ there is a set $Y \subset X$, $|Y| = r+s$, such that $Y^{(r)} \subset \mathcal{E} \cup \{F\}$.

Thus G is $(r+s)$-saturated if it contains no $K_{r+s}^{(r)}$ but if any edge (i.e. r-set) is added to G then the new graph does contain a $K_{r+s}^{(r)}$. In other words, G is $(r+s)$-saturated if it is a maximal graph without a $K_{r+s}^{(r)}$.

The Turán problem, discussed in §8, asks for the maximal number of edges in an $(r+s)$-saturated graph. Now we shall examine the minimum number of edges in an $(r+s)$-saturated graph.

Without much ado, let us construct an $(r+s)$-saturated r-graph which we shall show to be the unique extremal graph. Let $S_0 \subset X$, $|S_0| = s$, and set $G_0 = (X, \mathcal{E}_0)$ where

$$\mathcal{E}_0 = \{E \in X^{(r)} : E \cap S_0 \neq \emptyset\}.$$

Let us check that G_0 is $(r+s)$-saturated. Does it contain a $K_{r+s}^{(r)}$? Hardly, since if $Y \subset X$ and $|Y| = r+s$ then $|Y \setminus S_0| \geq r$ so Y has at least one r-set contained entirely in $Y \setminus S_0$ and so not belonging to $\mathcal{E}_0$. Shall we get a $K_{r+s}^{(r)}$ if we add an r-set to G_0? Yes, we shall, for if $F \in \mathcal{F}_0 = X^{(r)} \setminus \mathcal{E}_0$ then $F \cap S_0 = \emptyset$ so $Y = F \cup S_0$ has $r+s$ elements and Y has only one r-set not in $\mathcal{E}_0$, namely F. Hence $Y^{(r)} \subset \mathcal{E}_0 \cup \{F\}$.

The main result of this section, proved by Bollobás (1965), shows that G_0 is not only the unique $(r+s)$-saturated graph of minimal size but it is also the unique graph of minimal size in which the addition of any edge creates a new $K_{r+s}^{(r)}$.

Theorem 1. *Let $G = (X, \mathcal{E})$ be an r-graph in which the addition* *of any non-edge creates at least one new* $K^{(r)}_{r+s}$. *Then*

$$|\mathcal{E}| \geq \binom{n}{r} - \binom{n-s}{r},$$

with equality iff $G \cong G_0$. ∎

We shall deduce this result from another theorem, which looks rather technical but is, in fact, often very useful. Before we state and prove this technical theorem, we shall describe how an attempt to prove Theorem 1 leads us to Theorem 2. Suppose then that $G = (X, \mathcal{E})$ satisfies the conditions of Theorem 1, and write $\mathcal{F}$ for the set of non-edges: $\mathcal{F} = X^{(r)} \setminus \mathcal{E}$. Our aim is to show that $|\mathcal{F}|$ is bounded by some appropriate increasing function f of $|\mathcal{E}|$, namely that $|\mathcal{F}| \leq f(|\mathcal{E}|)$. As $|\mathcal{E}| + |\mathcal{F}| = \binom{n}{r}$, if we can choose f small enough, this will imply the theorem. We hope to achieve this by making each non-edge assign certain *weights* to the edges in such a way that each non-edge assigns at least a certain weight $w_0 > 0$ to the set of all edges, while no edge receives more than, say, weight 1 from the set of all non-edges. If this can be done then the total amount of weights assigned is at least $w_0|\mathcal{F}|$ and at most $|\mathcal{E}|$, so

$$|\mathcal{E}| \geq w_0|\mathcal{F}| = w_0\left(\binom{n}{r} - |\mathcal{E}|\right),$$

implying

$$|\mathcal{E}| \geq \frac{w_0}{1+w_0}\binom{n}{r}. \tag{1}$$

Hence if our weights can be chosen so that

$$|\mathcal{E}_0| = \frac{w_0}{1+w_0}\binom{n}{r}, \tag{2}$$

then the main part of Theorem 1 is proved.

Let us make the plan above a little more concrete. What we know about the r-graph G is that for every $F \in \mathcal{F}$ there are $(r+s)$-sets $Y_1, \ldots, Y_t$, $t = t(F) \geq 1$, such that $Y_i \supset F$ and $Y_i^{(r)} \cap \mathcal{F} = \{F\}$ for every i, $1 \leq i \leq t$. We hope to assign a certain positive weight $w(E, F)$ to all edges E in $\cup_{i=1}^t Y_i^{(r)}$ and weight $w(E, F) = 0$ to the other edges. Then the weight assigned by a non-edge $F \in \mathcal{F}$ to the set $\mathcal{E}$ of edges is

$$\sum_{E \in \mathcal{E}} w(E, F) \geq \sum_{E \in Y_1^{(r)} \setminus \{F\}} w(E, F). \tag{3}$$

As we do not want $w(E,F)$ to depend on G, but only on the relation between E and F, it has to be a function of $|E \cap F|$, say, in addition to the constants n, r and s. With such a choice, the right-hand-side of (3) is independent of F, say it is w_0. What else do we want from $w(E,F)$? That for every $E \in \mathcal{E}$, the total weight received by E from the set $\mathcal{F}$ of non-edges is at most 1:

$$\sum_{F \in \mathcal{F}} w(E,F) \leq 1. \tag{4}$$

Most importantly we would like to do this in such a way that in G_0 every edge receives precisely weight 1 from the set of non-edges, i.e. in such a way that in G_0 we have equality in (4).

If we want all these to be true then we no longer have any options: the structure of G_0 forces us to define $w(E,F)$ in a certain way. Indeed, in G_0 for every $F \in \mathcal{F}_0$ there is a unique set $Y_F \in X^{(r+s)}$ with $Y_F \supset F$ and $Y_F^{(r)} \cap \mathcal{F}_0 = \{F\}$. (In our earlier notation, $t = t(F) = 1$ and $Y_F = Y_1$.) Hence an edge $E \in \mathcal{E}_0$ receives some (non-zero) weight from a non-edge F iff $E \cap S_0 = E \cap F$. There are $\binom{n-s-|E\cap F|}{|F \setminus E|}$ such non-edges, all giving the same weight $w(E,F)$ to E. Hence we must have

$$w(E,F) = \binom{n-s-|E \cap F|}{|F \setminus E|}^{-1}.$$

How can we show that this choice of weights satisfies (4)? In other words, what will prevent us from assigning too much weight to an edge E of an $(r+s)$-saturated graph? Suppose F_1 and $F_2 \in \mathcal{F}$ both assign weights to $E \in \mathcal{E}_0$, i.e. $E \subset Y_1$ and $E \subset Y_2$, where $Y_i = Y_{F_i} : Y_i^{(r)} \subset \mathcal{E}_0 \cup \{F_i\}$, $i = 1,2$. From each F_i we define two disjoint subsets of the base set $X \setminus E$: $A_i = F_i \setminus E$ and $B_i = Y_i \setminus E \cup F_i$, $i = 1,2$. Then $A_1 \not\subset A_2 \cup B_2$ since otherwise $F_1 \subset A_1 \cup E \subset E \cup A_2 \cup B_2 \subset Y_2$, contradicting the fact that F_2 is the only non-edge in $Y_2^{(r)}$. Similarly $A_2 \not\subset A_1 \cup B_1$. What weight should the pair (A_1, B_1) get? The weight assigned by F_1 to E, namely

$$\binom{n-s-|E \cap F_1|}{|F_1 \setminus E|}^{-1} = \binom{n-r-|B_1|}{|A_1|}^{-1}.$$

Thus Theorem 1 will follow from the result below.

Theorem 2. *Let $A_i, B_i \subset X$, $i \in I$, be such that (i) $A_i \cap B_i = \emptyset$ and (ii) $A_i \not\subset A_j \cup B_j$ for $i,j \in I$, $i \neq j$. For $i \in I$ set $a_i = |A_i|$,*

$b_i = |B_i|$ and $w(i) = w(a_i, b_i, n) = \binom{n-b_i}{a_i}^{-1}$. Then

$$\sum_{i \in I} w(i) \leq 1 \tag{5}$$

with equality iff there is a set $B \subset X$ and an integer a, $1 \leq a \leq n - |B|$, such that $\{A_i : i \in I\} = (X \setminus B)^{(a)}$ and $B_i = B$ for every $i \in I$.

In particular, if $a_i = a$ and $b_i = b$ for all $i \in I$ then $|I| \leq \binom{n-b}{a}$.

Proof. Let us apply induction on n. For $n = 1$ the assertion is trivial so assume that $n > 1$ and the assertion holds for smaller values of n. We may also assume that $A_i \cup B_i \neq X$ for otherwise i is the only element of I and the assertion is trivial.

Let $X_1, \ldots, X_n$ be the $(n-1)$-element subsets of X. For $1 \leq j \leq n$ set

$$I_j = \{i \in I : A_i \subset X_j\}$$

and for $i \in I_j$ let $B_{ij} = B_i \cap X_j$, $b_{ij} = |B_i|$ and $w_j(i) = w(a_i, b_{ij}, n-1) = \binom{n-1-b_{ij}}{A_i}^{-1}$. Note that the system $\{A_i, B_{ij} : i \in I_j\}$ satisfies (i) and (ii) on the base set X_j. Hence, by the induction hypothesis,

$$\sum_{i \in I_j} w_j(i) \leq 1, \qquad j = 1, 2, \ldots, n. \tag{6}$$

There are $n - a_i$ subsets X_j that contain A_i, and $n - a_i - b_i$ of these contain B_i as well. Hence

$$\begin{aligned}\sum_{\{j : i \in I_j\}} w_j(i) &= (n - a_i - b_i)\binom{n-1-b_i}{a_i}^{-1} + b_i\binom{n-b_i}{a_i}^{-1} \\ &= n\binom{n-b_i}{a_i}^{-1} = nw(i).\end{aligned}$$

The second equality is true since

$$\begin{aligned}(n-b_i)\binom{n-b_i}{a_i}^{-1} &= \frac{(n-b_i)a_i!}{(n-b_i)\ldots(n-b_i-a_i+1)} \\ &= \frac{a_i!}{(n-b_i-1)\ldots(n-a_i-b_i+1)} \\ &= \frac{a_i!(n-a_i-b_i)}{(n-b_i-1)\ldots(n-a_i-b_i)} \\ &= (n-a_i-b_i)\binom{n-1-b_i}{a_i}^{-1}.\end{aligned}$$

Here the last step is valid because we have assumed that $n > a_i + b_i$.

Consequently, by (6),

$$\sum_{j=1}^{n} \sum_{i \in I_j} w_j(i) = \sum_{i \in I} \sum_{\{j : i \in I_j\}} w_j(i) = n \sum_{i \in I} w(i) \leq n,$$

proving (5).

It is readily seen that (5) is best possible: if $B \subset X$ and $1 \leq a \leq n - |B|$ then taking $\{A_i : i \in I\} = (X \setminus B)^{(a)}$ and $B_i = B$ for each $i \in I$, the conditions (i) and (ii) are satisfied and $w(i) = \binom{n-b}{a}^{-1}$ for every i. Since $|I| = \binom{n-b}{a}$, for this system we have equality in (5).

All that it remains to show are that these are the only systems for which we have equality in (5). This can be proved by induction on n, noting that if equality holds in (5) then we have equality in every inequality in the proof above. The details are left to the reader (Ex. 1). ∎

In view of the train of thought leading to Theorem 2, we have actually completed the proof of the inequality in Theorem 1. The uniqueness of the extremal graph is based on the characterization of the extremal set systems in Theorem 2 (Ex. 2). ∎

As we shall point out a little later, by making use of a reformulation of Theorem 2 and considering the complementary graph of a saturated graph, one can give a more direct proof of Theorem 1. But first let us note an immediate consequence of Theorem 1.

A set S is said to *represent* the sets in a set system $\mathcal{A}$ if $A \cap S \neq \emptyset$ for all $A \in \mathcal{A}$. The *transversal number* $\tau(\mathcal{A})$ of $\mathcal{A}$ is the minimal cardinality of a set representing $\mathcal{A}$. A set system $\mathcal{A}$ is said to be *τ-critical* if $\tau(\mathcal{A} \setminus \{A\}) < \tau(\mathcal{A})$ for all $A \in \mathcal{A}$. Thus an r-graph $G = (X, \mathcal{E})$ is τ-critical with $\tau(G) = \tau(\mathcal{E}) = t + 1$ if for every $Y \in X^{(t)}$ we have $(X \setminus Y)^{(r)} \cap \mathcal{E} \neq \emptyset$ but for every $E \in \mathcal{E}$ there is a set $Y \in X^{(t)}$ with $(X \setminus Y)^{(r)} \cap \mathcal{E} = \{E\}$. Hence an r-graph $G = (X, \mathcal{E})$ is τ-critical with $\tau(\mathcal{E}) = t + 1$ iff the complementary r-graph $\overline{G} = (X \setminus X^{(r)}, \mathcal{E})$ is $(n - t)$-saturated, where $n = |X|$. Thus, as noted by Bollobás (1965), Theorem 1 has the following consequence.

Corollary 3. *Let $G = (X, \mathcal{E})$ be a τ-critical r-graph with $\tau(G) = t + 1$. Then*

$$|\mathcal{E}| \leq \binom{r+t}{r}$$

with equality iff $\mathcal{E} = Y^{(r)}$ for some set Y with $|Y| = r + t$, i.e. iff G is the union of $K_{r+t}^{(r)}$ and isolated vertices. ■

In the original paper (Bollobás (1965)) Theorem 2 appears as a lemma since its statement is not too attractive and, as above, it was used to prove Theorem 1. Because of its numerous applications, in the presentation above we upgraded it to a theorem. In fact, Theorem 2 is not as esoteric as it seems at first glance — just the contrary, it belongs to the mainstream of the theory of set systems. It not only contains Sperner's theorem but it is also a considerable extension of the celebrated LYM inequality! Indeed, on putting $B_i = \emptyset$ for all i in Theorem 2 the condition becomes that $\{A_i : i \in I\}$ is a Sperner system and we obtain precisely the LYM inequality, Theorem 3.2.

It is not surprising that Lubell's proof of the LYM inequality, given in §3, can be boosted to a proof of inequality (5) in the theorem of Bollobás. Before giving this proof let us note that by writing B_i for the set $X \setminus A_i \cup B_i$, we obtain a reformulation of the theorem in which the sets A_i and B_i play similar roles and, as in Corollary 3, the size of the ground set is irrelevant.

Theorem 2′. *For two non-negative integers a and b write $w(a, b) = \binom{a+b}{a}^{-1} = \binom{a+b}{b}^{-1}$. Let $\{(A_i, B_i) : i \in I\}$ be a finite collection of finite sets such that $A_i \cap B_j = \emptyset$ iff $i = j$. For $i \in I$ set $a_i = |A_i|$, $b_i = |B_i|$. Then*

$$\sum_{i \in I} w(a_i, b_i) \leq 1 \tag{5'}$$

with equality iff there is a set Y and integers a, b such that $0 \leq a \leq a + b = |Y|$ and $\{(A_i, B_i) : i \in I\}$ is the collection of all ordered pairs of disjoint subsets of Y with $|A_i| = a$ and $|B_i| = b$.

In particular, if $a_i = a$ and $b_i = b$ for all $i \in I$ then $|I| \leq \binom{a+b}{a}$. ■

Let us see then the promised proof of inequality (5′). Assume that $A_i \cup B_i$ are subsets of a set X. We shall say that a permutation $x_1 x_2 \ldots x_n$ of X is *compatible* with an ordered pair (A, B) of disjoint subsets of X if no element of B precedes an element of A, i.e. if $x_i \in A$ and $x_j \in B$ imply $i < j$. If $|A| = a$ and $|B| = b$ and A and B are disjoint then they are compatible with

$$\binom{n}{a+b} a!\, b!\, (n - a - b)! = w(a, b) n!$$

permutations. Here $\binom{n}{a+b}$ counts the number of choices for the positions of $A \cup B$ in the permutations; having chosen these positions, A has to occupy the first a places, giving $a!$ choices for the order of A and $b!$ choices for the order of B; the remaining elements can be chosen in $(n-a-b)!$ orders.

Each of the $n!$ permutations is compatible with at most one pair (A_i, B_i), $i \in I$. Indeed, if $x_1x_2\dots x_n$ were compatible with (A_i, B_i) and (A_j, B_j) and, say, we had $\max\{k : x_k \in A_i\} \le \max\{k : x_k \in A_j\}$ then we would have $\min\{k : x_k \in B_j\} > \max\{k : x_k \in A_j\} \ge \max\{k : x_k \in A_i\}$ and so $A_i \cap B_j = \emptyset$, contradicting our assumption.

Putting these two bounds together, we find that

$$\sum_{i \in I} w(a_i, b_i)n! \le n!,$$

which is precisely (5′).

There is nothing magic in the formulation Theorem 2′ rather than Theorem 2, and the use of permutations rather than maximal chains in $\mathcal{P}(X)$. Indeed, a permutation $\pi = x_1x_2\dots x_n$ corresponds to the maximal chain $\emptyset \subset \{x_1\} \subset \{x_1, x_2\} \subset \dots \subset \{x_1, \dots, x_n\} = X$ and in the notation of Theorem 2 compatibility means that the maximal chain intersects the interval $\mathcal{J}_i = [A_i,\ X \setminus B_i] = \{C :\ A_i \subset C \subset X \setminus B_i\}$. It is easily checked that $w(a_i, b_i;\ n)$ maximal chains intersect an interval $\mathcal{J}_i$ and each maximal chain intersects at most one interval $\mathcal{J}_i$ (see Ex. 3).

Let us point out that Theorem 2′ can be used to give more direct proofs of Theorem 1 and Corollary 3. Let us see first Corollary 3. Let $G = (X,\ \mathcal{F})$ be a τ-critical r-graph with $\tau(\mathcal{F}) = t + 1$. This implies that if $\mathcal{F} = \{A_1, A_2, \dots, A_m\}$ then for every A_i there is a set $B_i \subset X$, $|B_i| = t$, such that B_i is disjoint from A_i but meets every other edge A_j. Hence, by Theorem 2′, we have $m \le \binom{r+t}{r}$.

Theorem 1 can be proved in a similar way. Let $G = (X, \mathcal{E})$ be an r-graph with $\mathcal{F} = X^{(r)} \setminus \mathcal{E} = \{A_1, A_2, \dots, A_m\}$. Suppose for every $A_i \in \mathcal{F}$ there is a set $S_i \subset X$, $|S_i| = s$, such that $A_i \cap S_i = \emptyset$ and $A_j \not\subset A_i \cup S_i$ for $j \ne i$. Set $B_i = X \setminus A_i \cup S_i$ so that $|A_i| = r$, $|B_i| = n - r - s$ and $A_i \cap B_j = \emptyset$ iff $i = j$. Hence, by Theorem 2′, $m \le \binom{n-s}{r}$ so $|\mathcal{E}| \ge \binom{n}{r} - \binom{n-s}{r}$. By simply rewriting the above proof of (5′) word for word, we obtain a generalization of Theorem 2′ (to be precise, inequality (5′)) from pairs of sets to k-tuples of sets, though this generalization is only for its own sake.

Theorem 4. *Let $\{(A_{i1}, A_{i2}, \ldots, A_{ik}) : i \in I\}$ be a family of k-tuples of pairwise disjoint finite sets such that for all $h, i \in I$, $h \neq i$ there is an integer l, $1 \leq l < k$, such that*

$$(\bigcup_{j=1}^{l} A_{hj}) \cap (\bigcup_{j=l+1}^{k} A_{ij}) \neq \emptyset$$

and

$$(\bigcup_{j=1}^{l} A_{ij}) \cap (\bigcup_{j=l+1}^{k} A_{hj}) \neq \emptyset.$$

Then

$$\sum_{i \in I} w(|A_{i1}|, \ldots, |A_{ik}|) \leq 1$$

where

$$w(a_1, \ldots, a_k) = \binom{a_1 + \ldots + a_k}{a_1, \ldots, a_k}^{-1}. \qquad \blacksquare$$

This result is best possible as the following obvious example shows. Given $a_1, a_2, \ldots, a_k$, let A be a set with $a_1 + \ldots + a_k$ elements and take all partitions of A of the form

$$A = \bigcup_{j=1}^{k} A_{ij} \qquad \text{where } |A_{ij}| = a_j,\ 1 \leq j \leq k.$$

Theorem 4 has a rather natural reformulation in terms of unrelated chains, though with an artificial weight function. Given a chain $E_1 \subset E_2 \subset \ldots \subset E_k$, define the *dual chain* as $E'_1 \supset E'_2 \supset \ldots \supset E'_k = \emptyset$ where $E'_i = E_k \setminus E_i$. Call two chains $(E_i)_1^k$ and $(F_i)_1^k$ *unrelated* if there is an integer l, $1 \leq l < k$, such that

$$E_l \cap F'_l \neq \emptyset \qquad \text{and} \qquad E'_l \cap F_l \neq \emptyset$$

where $(E'_i)_1^k$ and $(F'_i)_1^k$ are the dual chains. Define the *weight* of a chain $C = (E_i)_1^k$ as

$$w(C) = w(|E_1|, |E_2 \setminus E_1|, \ldots, |E_k \setminus E_{k-1}|)$$

where

$$w(a_1, \ldots, a_k) = \binom{a_1 + \ldots + a_k}{a_1, \ldots, a_k}^{-1}.$$

Theorem 4′. *Let C_i, $i \in I$, be a family of unrelated chains of k sets each. Then*

$$\sum_{i\in I} w(C_i) \le 1. \qquad \blacksquare$$

To recover the LYM inequality, take chains of the form $E_1 \subset E_2 = X$. Then two chains $E_1 \subset E_2 = X$ and $F_1 \subset F_2 = X$ are unrelated iff $E_1 \not\subset F_1$ and $F_1 \not\subset E_1$. Theorem 2 of Bollobás is the case $k = 2$ of Theorems 4 and 4′.

Exercises

1. Check that the proof of Theorem 2 implies that equality holds in (5) iff $\{A_i : i \in I\} = (X \setminus B)^{(a)}$ for some $B \subset X$ and $1 \le a \le n - |B|$.

2. Use the characterization of extremal systems in Theorem 2 to prove the characterization of extremal configurations in Theorem 1.

3. Rewrite the proof of (5′) in terms of maximal chains and intervals $[A_i, X \setminus B_i]$ to give a detailed proof of inequality (5) in Theorem 2.

4. Check that the proof of (5′) extends to a proof of Theorem 4.

5. Let $A_{i1} \subset A_{i2}$, $i \in I$, be chains of subsets of X such that $|A_{i2}| \ge |A_{i1}| + k$ and $A_{i1} \not\subset A_{j2}$ if $i \ne j$. Prove that

$$|I| \le \binom{n-k}{\lfloor (n-k)/2 \rfloor}.$$

Show also that equality holds iff there is a set $B \in X^{(k)}$ such that

$$\{A_{i1} : i \in I\} = (X \setminus B)^{(l)}$$

and $A_{i2} = A_{i1} \cup B$ for all $i \in I$, where l is $\lfloor (n-k)/2 \rfloor$ or $\lceil (n-k)/2 \rceil$.

6. Note the following trivial extension of Corollary 3: if a set system $\mathcal{A}$ is p-critical then

$$\sum_{A\in\mathcal{A}} \binom{|A|+p-1}{p-1}^{-1} \le 1.$$

(Tuza (1984)).

7. Denote by $f_k(n)$ the maximal value of m for which there are m chains $A_{i0} \subset A_{i1} \subset \ldots \subset A_{ik} \subset X$, $A_{ij} \neq A_{ik}$ for $j \neq k$, $i = 1, 2, \ldots, m$, such that

$$A_{ij} \not\subset A_{i'h} \qquad \text{if } i \neq i'.$$

Deduce from Exercise 5 that

$$f_k(n) = \binom{n-k}{\lfloor (n-k)/2 \rfloor}.$$

(Griggs, Stahl and Trotter (1984)).

8. A poset $P = (S, \leq)$ can be embedded in a poset $P' = (S', \leq)$ if there is an order preserving injection $\varphi : P \to P'$ (thus $x < y$ iff $\varphi(x) < \varphi(y)$). Let P_l be the linearly ordered poset on l elements and for an integer m and a poset P, let mP be the disjoint union of m copies of P. The 2-*dimension* of a poset P, denoted by $\dim_2(P)$, is the minimal n for which P can be embedded in the poset $\mathcal{P}([n])$.

Deduce from Exercise 7 that for $k \geq 0$ and $m \geq 1$ we have

$$\dim_2(mP_{k+1}) = \min\left\{n : \binom{n-k}{\lfloor (n-k)/2 \rfloor} \geq m\right\}.$$

(Griggs, Stahl and Trotter (1984))

9. (A little long.) For a natural number u, let $b(u) = \binom{u}{\lfloor u/2 \rfloor}$. Let $m = b(n)$, $I = [m]$ and let $\Delta = \{(i,i) : i \in I\}$ be the diagonal of $I \times I$. Let $U_k, V_k \subset Y$, $k = 1, \ldots, n$, be such that

$$I \times I \setminus \Delta = \bigcup_{k=1}^{n} U_k \times V_k.$$

Prove that

$$\sum_{k=1}^{n} \{|U_k| + |V_k|\} \geq nm.$$

Show also that equality can hold for all n. (Tarján (1975)).
(For $1 \leq i \leq m$ let $A_i = \{k : y_i \in U_k\}$ and $B_i = \{k : y_i \in V_k\}$. Note that $\{(A_i, B_i) : i \in I\}$ satisfies the conditions of Theorem 2′ and that $\sum_{k=1}^{n}\{|U_k| + |V_k|\} = \sum_{i=1}^{n}\{|A_i| + |B_i|\}$. Conclude that the sum in question is at least as large as $\sum_{i=1}^{m} c_i$ where $\sum_{i=1}^{m} b(c_i)^{-1} \leq 1$. Finally, show that if $2 \leq c \leq d-2$ then

$$b(c+1)^{-1} + b(d-1)^{-1} < b(c)^{-1} + b(d)^{-1}$$

and complete the proof.)

§10. WELL-SEPARATED SYSTEMS

Given k and n, what is the largest m for which we can find m subsets of a set of order n such that the symmetric difference of any two of these subsets has at least k elements? In a slightly different formulation, this is precisely one of the basic questions in coding theory: how many $0,1$ sequences of length n can we construct if any two of the sequences must differ in at least k places? Our main aim in this section is to present a result of Corrádi and Kátai (1969), concerning the case $k = \lceil n/2 \rceil$. The result is particularly fascinating because the bound it gives, which is essentially best possible, depends considerably on the form of n. Most of the work in proving the main assertion will be put into the following lemma about partitioning vectors in $\Re^n$, the n-dimensional Euclidean space. As usual, (x, y) denotes the standard inner product of x and $y \in \Re^n$: if $x = (x_1, x_2, \ldots, x_n)$ and $y = (y_1, y_2, \ldots, y_n)$ then $(x, y) = \sum_{i=1}^n x_i y_i$.

Lemma 1. *Let $n \geq 1$, $r \geq 1$ and let $S = \{x_1, x_2, \ldots, x_{n+r}\}$ be a set of $n + r$ non-zero vectors in $\Re^n$ such that*

$$(x_i, x_j) \leq 0 \text{ whenever } 1 \leq i < j \leq n + r.$$

(i) Under these conditions, $r \leq n$, and if $r = n$ then $\Re^n$ has an orthonormal basis $\{a_1, \ldots, a_n\}$ such that, for all $1 \leq i \leq n$, $S = \{\lambda_1 a_1, \mu_1 a_1, \lambda_2 a_2, \mu_2 a_2, \ldots, \lambda_n a_n, \mu_n a_n\}$ where $\mu_i < 0 < \lambda_i$.

(ii) The set S can be partitioned into r non-empty sets: $S = \bigcup_{i=1}^r S_i$ such that $(x, y) = 0$ whenever $x \in S_i$, $y \in S_j$ and $1 \leq i < j \leq r$.

Proof. (i) Let us apply induction on n. For $n = 1$ the assertion being trivial, let $n \geq 2$ and suppose the result is true for smaller values of n. Let us assume also that $r \geq n$; our aim is then to show that $r = n$ and that S is of the required form.

Let $a_1 = x_1/|x_1|$ be the unit vector in the direction of x_1, let $H \simeq \Re^{n-1}$ be the hyperplane orthogonal to a_1 and let Q be the orthogonal projection onto H. Thus

$$H = \{x \in \Re^n : Qx = x\} = \{x \in \Re^n : (x, a_1) = 0\}$$

and

$$Qx = x - (x, a_1)a_1$$

for every $x \in \Re^n$. Set $y_1 = Qx_i$, $i = 2, 3, \ldots, n + r$. Note that for $2 \leq i < j \leq n + r$ we have

$$\begin{aligned}(y_i, y_j) &= (x_i - (x_i, a_1)a_1,\ x_j - (x_j, a_1)a_1) \\ &= (x_i, x_j) - (x_i, a_1)(x_j, a_1) \leq (x_i, x_j) \leq 0.\end{aligned} \tag{1}$$

Furthermore, $(y_i, y_j) = 0$ implies that $(x_i, x_j) = 0$ and at least one of x_i and x_j belongs to H.

Since $n + r - 1 \geq 2n - 1 > 2(n - 1)$, relation (1) and the induction hypothesis imply that at least one of the y_i's is 0, say $y_2 = 0$. But then x_2 is a negative multiple of a_1 and so $(x_i, a_1) = 0$ for every i, $3 \leq i \leq n + r$, i.e., $x_3, x_4, \ldots, x_{n+r}$ are $n + r - 2 \geq 2(n - 1)$ distinct nonzero vectors in $H \simeq \Re^{n-1}$. Once again, by the induction hypothesis, $r = n$ and H has an orthonormal basis $a_2, a_3, \ldots, a_n$ such that $\{x_3, \ldots, x_{2n}\} = \{\lambda_2 a_2, \mu_2 a_2, \lambda_3 a_3, \mu_3 a_3, \ldots, \lambda_n a_n, \mu_n a_n\}$ where $\mu_i < 0 < \lambda_i$, $2 \leq i \leq n$. Hence S has the required form.

(ii) Let us apply induction on n. By part (i), we must have $n \geq r$, and for $n = r$ a considerably stronger assertion holds than we claim here. Hence we may assume that $n \geq r + 1$ and the assertion holds for smaller values of n.

Consider $x_1 \in S$, set $a_1 = x_1/|x_1|$, and let H, Q and $y_2, y_3, \ldots, y_{n+r}$ be as in (i). Set $\tilde{S} = \{y_2, y_{n+r}\} \subset H$.

If $0 \in \tilde{S}$, say $y_2 = 0$, the proof can be completed as in (i) since then $x_3 = y_3$, $x_4 = y_4, \ldots$, $x_{n+r} = y_{n+r}$ are $n + r - 2 = n - 1 + (r - 1)$ vectors in the $(n - 1)$-dimensional subspace H and the subspace H is orthogonal to x_1 and x_2.

Suppose then that $0 \notin \tilde{S}$. Since $(y_i, y_j) \leq 0$ for $2 \leq i < j \leq n + r$, the $(n-1) + r \leq 2(n-1)$ vectors $y_2, y_3, \ldots, y_{n+r}$ can be partitioned into r nonempty orthogonal sets: $\tilde{S} = \bigcup_{i=1}^{r} \tilde{S}_i$. Set

$$S_i = \{x_j : y_j \in \tilde{S}_i\}, \qquad i = 1, 2, \ldots, r.$$

We claim that at most one of the r sets $S_1, S_2, \ldots, S_r$ is not contained in H, i.e. there is at most one S_i with $S_i \neq \tilde{S}_i$. Indeed, if $S_k \not\subset H$

and $S_l \not\subset H$ for some $1 \le k < l \le r$ then there are $x_i \in S_k$ and $x_j \in S_l$ such that $(x_1, x_i) < 0$ and $(x_1, x_j) < 0$. But this is impossible since then, as we remarked after inequality (1), we have $(Qx_i, Qx_j) = (y_i, y_j) < 0$, contradicting the orthogonality of $\tilde{S}_k$ and $\tilde{S}_l$.

The proof is almost complete. We may assume that $S_2, S_3, \ldots, S_r$ are contained in H and so $S_2 = \tilde{S}_2$, $S_3 = \tilde{S}_3, \ldots$, $S_r = \tilde{S}_r$. Hence $S_2, S_3, \ldots, S_r$ are orthogonal to each other, to $\tilde{S}_1$ and to x_1. But $S_1 \cup \{x_1\}$ is contained in the linear span of $\tilde{S}_1$ and x_1, so each S_i, $2 \le i \le r$, is orthogonal to $S_1 \cup \{x_1\}$ as well. Hence

$$S = (S_1 \cup \{x_1\}) \cup \bigcup_{i=2}^{r} S_i$$

is an appropriate partition. ∎

Let us call a system $\mathcal{F} \subset \mathcal{P}(X)$ *well-separated* if for all $F_1, F_2 \in \mathcal{F}$, $F_1 \ne F_2$, $|F_1 \triangle F_2| = |(F_1 \cap \overline{F}_2) \cup (\overline{F}_1 \cup F_2)| \ge n/2$ where, as always, $X = [n] = \{1, 2, \ldots, n\}$, and $\overline{F}_i = X \setminus F_i$ is the complement of $F_i \subset X$. Corrádi and Kátai (1969) proved the following bounds on the size of a well-separated system.

Theorem 2. *Let $\mathcal{F} \subset \mathcal{P}(X)$ be a well-separated system. Then*

$$|\mathcal{F}| \le n+1 \qquad \text{if } n \text{ is odd} \tag{2}$$

$$|\mathcal{F}| \le n+2 \qquad \text{if } n \equiv 2 \pmod 4 \tag{3}$$

and

$$|\mathcal{F}| \le 2n \qquad \text{if } n \equiv 0 \pmod 4. \tag{4}$$

Proof. Let $\mathcal{F} = \{F_1, F_2, \ldots, F_{n+r}\}$ and for $1 \le i \le n+r$ set $f_i = (f_{i1}, f_{i2}, \ldots, f_{in}) \in \Re^n$ where

$$f_{ik} = \begin{cases} 1 & \text{if } k \in F_i, \\ -1 & \text{if } k \notin F_i. \end{cases}$$

Then for $1 \le i < j \le n+r$,

$$\begin{aligned}(f_i, f_j) &= |\{k : f_{ik} = f_{jk}\}| - |\{k : f_{ik} \ne f_{jk}\}| \\ &= |(F_i \cap F_j) \cup (\overline{F}_i \cap \overline{F}_j)| - |(F_i \cap \overline{F}_j) \cup (\overline{F}_i \cap F_j)| \\ &= n - 2|F_i \triangle F_j| \le 0.\end{aligned}$$

Consequently, by Lemma 1, $r \le n$, i.e. $|\mathcal{F}| \le 2n$, proving (4).

Our next aim is to prove (2). Suppose that $r \ge 2$. Then, by Lemma 1, the vectors $f_1, f_2, \ldots, f_{n+r}$ can be split into $r \ge 2$ non-empty classes which are orthogonal to each other. In particular, some two of these vectors are orthogonal, say $(f_i, f_j) = 0$. But then $(f_i, f_j) = n - 2|F_i \triangle F_j| = 0$ so n is even, proving (2).

In order to prove (3), suppose that $r \ge 3$. Then, again by Lemma 1, our vectors can be split into three orthogonal classes, so some three of them are pairwisely orthogonal, say, $(f_i, f_j) = (f_j, f_k) = (f_i, f_k) = 0$, where $1 \le i < j < k \le n + r$. Thus $|F_i \triangle F_j| = |F_j \triangle F_k| = |F_i \triangle F_k| = n/2$. Furthermore, $|\overline{F}_j \triangle F_k| = |X \setminus (F_j \triangle F_k)| = n - |F_j \triangle F_k| = n/2$, similarly $|\overline{F}_i \triangle F_k|$ and so

$$|F_i \triangle F_j| + |\overline{F}_j \triangle F_k| - |\overline{F}_i \triangle F_k| = n/2. \tag{5}$$

The left-hand-side of (5) is precisely

$$2|F_i \cap \overline{F}_j \cap \overline{F}_k| + 2|\overline{F}_i \cap F_j \cap F_k|,$$

so n is a multiple of 4. This proves (3). ■

Can we have equality in each of the relations (2), (3) and (4) for infinitely many values of n? Let us consider the most interesting case, that of (4). If $\mathcal{F} = \{F_1, F_2, \ldots, F_{2n}\}$ is a well-separated system then, by Lemma 1, we may assume that $f_{n+i} = -f_i$ for $1 \le i \le n$, and the vectors $f_1, f_2, \ldots, f_n$ are orthogonal to each other. Hence if A is the matrix whose ith row is f_i, then A is an n by n matrix of $+1$'s and -1's, whose rows are orthogonal to each other (and hence, so are the columns). Such a matrix is said to be an *Hadamard matrix*. Conversely, if A is an n by n Hadamard matrix, let $f_i = (f_{i1}, f_{i2}, \ldots, f_{in})$ be the ith row and set $f_{n+i} = -f_i$, $i = 1, 2, \ldots, n$, and $F_i = \{j \in X : f_{ij} = 1\}$, $i = 1, 2, \ldots, 2n$. Since $(f_i, f_j) = 0$ if $1 \le i < j \le 2n$ and $j \ne n + i$, we have $|F_i \triangle F_{n+i}| = n$ if $1 \le i \le n$ and $|F_i \triangle F_j| = n/2$ if $1 \le i < j \le 2n$ and $j \ne n + i$. Hence

$$\mathcal{F}(A) = \{F_i : 1 \le i \le 2n\}$$

is a well-separated family.

There are many ways of constructing Hadamard matrices. The one by one matrices (1) and (-1) are, of course, Hadamard matrices, and if A is an n by n Hadamard matrix then the $2n$ by $2n$ matrix

$$\begin{pmatrix} A & A \\ A & -A \end{pmatrix}$$

is clearly an Hadamard matrix. Hence there is an Hadamard matrix of order 2^n for every $n \geq 1$.

Paley (1933) used quadratic residues to construct Hadamard matrices which, by now, have come to be called *Paley matrices*. Let q be a prime congruent to 3 modulo 4 and let $\mathbf{F}_q$ be the field of order q, i.e. the field of integers modulo q. Let $P_q = (p_{ij})$ be the q by q matrix with $p_{ij} = 1$ if $i \neq j$ and $j - i$ is a square (i.e. a quadratic residue) in $\mathbf{F}_q$, and $ij = -1$ otherwise.

Since $q \equiv 3 \pmod 4$, -1 is not a square in $\mathbf{F}_q$, so $1 \leq a \leq q-1$ is a square iff $q - a$ is not a square. In particular, each row has $(q+1)/2$ -1's and $p_{ij} = -p_{ji}$ for $i \neq j$. Thus P_q is a skew-symmetric matrix with -1's in the main diagonal and satisfying $p_{i+1,j+1} = p_{i,j}$ for $1 \leq i \leq q-1$ and $1 \leq j \leq q-1$.

Let A_{q+1} be the $(q+1)$ by $(q+1)$ matrix obtained from P_q by adding to it a $(q+1)$st row and a $(q+1)$st column, with all new entries being 1's. As every row of the Paley matrix P_q has precisely $(q+1)/2$ entries which are -1's, and the product of any two rows is -1, the matrix A_{q+1} is an Hadamard matrix (see Fig. 1).

$$\begin{pmatrix}
- & + & + & - & + & - & - & + \\
- & - & + & + & - & + & - & + \\
- & - & - & + & + & - & + & + \\
+ & - & - & - & + & + & - & + \\
- & + & - & - & - & + & + & + \\
+ & - & + & - & - & - & + & + \\
+ & + & - & + & - & - & - & + \\
+ & + & + & + & + & + & + & +
\end{pmatrix}$$

Figure 1. *The Hadamard matrix* A_8, *with* $+$ *and* $-$ *representing* 1 *and* -1.

The same construction can be carried out for every q which is a prime power congruent to 3 modulo 4. Colouring this with the "doubling operation" described earlier, we see that there is a Hadamard matrix of order n whenever $n = 2^r(q+1)$, where $q \equiv 3 \pmod 4$ is a prime power.

A matrix obtained from an Hadamard matrix by multiplying a row or a column by -1 is again a Hadamard matrix. By a sequence of such operations we can change an Hadamard matrix into one in which every entry in the last row is 1 and so is every entry in the last column. Given such an m by m matrix A, let $\overline{A}$ be the m by $m-1$ matrix obtained from A by deleting the last column, and let $\overline{\overline{A}}$ be the m by $m-2$ matrix obtained from A by deleting the last two columns. Then the inner product of any two rows of $\overline{A}$ is -1 and the inner product of any two rows of $\overline{\overline{A}}$ is 0 or -2. Hence, for $n = m-1$, the matrix $\overline{A}$ gives a well-separated set system with $m = n+1$ elements (in fact, with $|F_i \triangle F_j| = (n+1)/2$ for all $i \neq j$) and for $n = m-2$, the matrix $\overline{\overline{A}}$ gives a well-separated set system with $m = n-2$ elements. (Note that from the matrix A_8 in Figure 1 we obtain the set systems $\{235, 346, 457, 156, 267, 137, 124, 1234567\} \subset \mathcal{P}([7])$ and $\{235, 346, 45, 156, 26, 13, 124, 123456\} \subset \mathcal{P}([6])$.) Hence equality is attained in (2) whenever there is a Hadamard matrix of order $n+1$, and equality is attained in (3) whenever there is a Hadamard matrix of order $n+2$.

We summarize our observations in the following theorem.

Theorem 3. *(i) A set system $\mathcal{F} \subset \mathcal{P}(X)$ consisting of $2n$ sets is a well-separated family iff $\mathcal{F} = \mathcal{F}(A)$ for some Hadamard matrix of order n.*

(ii) Let $H = \{m \in \mathbf{N}$: there is a Hadamard matrix of order $m\}$. Then $H \supset \{2^r(q+1) : r \geq 0, q \equiv 3$ is a prime power $\}$.

(iii) If $m \in H$, $m \geq 4$, then in (4), equality can be attained for $n = m$, in (3), equality can be attained for $n = m-2$ and in (2), equality can be attained for $n = m-1$. ∎

For considerably more information about Hadamard matrices and the set H, the reader is advised to consult Hall (1967).

Let us return to the general problem of finding large set systems $\mathcal{F}$ whose sets differ in many elements, say

$$d(A, B) = |A \triangle B| \geq k \text{ for all } A, B \in \mathcal{F},\ A \neq B.$$

So far we have studied the case $k = n/2$; now let us consider the case $k = \alpha n$ where $0 < \alpha < 1$ is fixed. Define

$$f(n; \alpha) = \\ = \max\{|\mathcal{F}| : \mathcal{F} \subset \mathcal{P}(X), d(A, B) \geq \alpha n \text{ for all } A, B \in \mathcal{F}, A \neq B\}.$$

In Theorem 2 we gave an essentially best possible upper bound for $f(n; 1/2)$. The result is particularly fascinating because it shows that $f(n; \alpha)$ depends heavily on the form of n, it is not nearly as smooth as we would expect it.

In fact, as we shall see, in a certain sense, the function $f(n; \alpha)$ is most interesting when $\alpha = 1/2$. In the proof of Theorem 2 we made use of the fact that $d(F_i, F_j) = |F_i \triangle F_j|$ is related to the inner product (f_i, f_j) of the corresponding vectors:

$$(f_i, f_j) = n - 2|F_i \triangle F_j|.$$

Hence to get a set system $\mathcal{F} = \{F_1, \ldots, F_m\}$ with $d(A, B) = |A \triangle B| \geq \alpha n$, we must find vectors $f_1, \ldots, f_m \in \Re^n$, with coordinates 1 and -1, such that $(f_i, f_j) \leq n - 2\alpha n$.

Normalizing the vectors, i.e. taking the unit vector $u_i = f_i/|f_i| = f_i/\sqrt{n}$ instead of f_i, we are led to the problem of finding many unit vectors $u_1, \ldots, u_m$ in $\Re^n$ such that the angle between any two of them is large, i.e. $(u_i, u_j) \leq 1 - 2\alpha = \delta$ for all $i \neq j$. The reader may find it surprising that for a fixed negative number δ the number of u_i's we can find in $\Re^n$ is bounded by a number independent of n.

Lemma 4. *Let $u_1, \ldots, u_m$, $m \geq 2$, be unit vectors in $\Re^n$ with*

$$(u_i, u_j) \leq \delta < 0 \qquad \textit{for all } i, j;\ 1 \leq i < j \leq m.$$

Then

$$m \leq 1 - \frac{1}{\delta}$$

and if equality holds then $\sum_1^m u_i = 0$ and $(u_i, u_j) = \delta$ for all i, j, $1 \leq i < j \leq m$.

Proof. As

$$0 \leq \Big|\sum_1^m u_i\Big|^2 = \Big(\sum_1^m u_i, \sum_1^m u_i\Big) = \sum_1^m |u_i|^2 + 2\sum_{i<j}(u_i, u_j)$$
$$\leq m + m(m-1)\delta,$$

the result follows. ∎

Lemma 4 implies the following result about set systems with large symmetric differences.

Lemma 5. *Let $F_1, \ldots, F_m \subset X$ be such that*

$$d(F_i, F_j) = |F_i \triangle F_j| \geq d > n/2.$$

Then

$$m \leq \frac{2d}{2d - n},$$

and if equality holds then every $x \in X$ belongs to precisely $m/2$ sets F_i and $|F_i \triangle F_j| = d$ for all i, j, $1 \leq i < j \leq m$. ■

The results above prove two of the three assertions below showing that $f(n; \alpha)$ has a double jump at $\alpha = 1/2$. We leave the proof of the third to the reader (Ex. 4).

Theorem 6. *Let $0 < \alpha < 1$ be fixed. If $\alpha > 1/2$ then*

$$f(n; \alpha) \leq \frac{2\alpha}{2\alpha - 1}$$

if $\alpha < 1/2$ then there is an $\epsilon = \epsilon(\alpha) > 0$ such that

$$f(n; \alpha) \geq 2^{\epsilon n}$$

and if $\alpha = 1/2$ then

$$f(n; 1/2) \leq 2n$$

with equality for infinitely many values of n. ■

It is interesting to note that Lemma 5 also implies the first part of Theorem 2, namely inequality (2). Indeed, if $n = 2k + 1$ and $\mathcal{F}$ is well separated then $|F_i \triangle F_j| \geq k + 1$ for all $i \neq j$ so

$$|\mathcal{F}| \leq \frac{2(k+1)}{2(k+1) - (2k+1)} = 2k + 2 = n + 1.$$

How large a set system $\mathcal{F} = \{F_1, \dots, f_m\}$ can we obtain if instead of demanding $d(F_i, F_j) \geq k$, we demand that $d(F_i, F_j) = k$ for all $i \neq j$? We have seen that if there is an Hadamard matrix of order $h \geq 4$ then for $n = h - 1$ and $k = \frac{h}{2} = \frac{n+1}{2}$ we can have $m = n + 1$. The next, rather simple, result shows that this is the best we can do.

Theorem 7. *Suppose $\mathcal{F} = \{F_1, \dots, F_m\}$ is a set system on X such that $d(F_i F_j) = k$ for all $i \neq j$. If $k = \frac{n+1}{2}$ then $m \leq n + 1$, otherwise $m \leq n$.*

Proof. The assertion is trivial if either $n = 1$ or $k = 0$. Assume therefore that $n \geq 2$ and $k \geq 1$. Define $f_1, \dots, f_m \in \Re^n$ as in the proof of Theorem 2. Then $|f_i|^2 = n$ and $(f_i, f_j) = n - 2k = a$, say, for all $i \neq j$.

If $f_1, \ldots, f_m$ are linearly independent then $m \leq n$ so we are done. Suppose then that these vectors are not linearly independent, say

$$\sum_{i=1}^{m} \xi_i f_i = 0 \qquad \text{and} \qquad \sum_{i=1}^{m} |\xi_i| > 0.$$

Then

$$0 = \sum_{i=1}^{m} \xi_i (f_i, f_j) = a \sum_{i=1}^{m} \xi_i + (n-a)\xi_j$$

so

$$\xi_j = \frac{a}{a-n} \sum_{i=1}^{m} \xi_i.$$

Therefore $\sum_{i=1}^{m} \xi_i \neq 0$ and

$$\sum_{j=1}^{m} \xi_j = \frac{ma}{a-n} \sum_{i=1}^{m} \xi_i,$$

implying $ma = a - n$ and so $m = (a-n)/a$. Hence either $m \leq n$ or else $a = -1$ and $m = n+1$. If $a = -1$ then $k = (n+1)/2$. ∎

Recently Frankl and Rödl (1986) proved a deep and difficult theorem related to the results above. (A weaker form of this theorem was conjectured by Larman and Rogers (1972).) For every $r \geq 2$ there is an $\epsilon_r > 0$ with the following property. If $2^{4n} \geq r$ and $\mathcal{F} \subset \mathcal{P}([4n])$ does not contain r sets $F_1, \ldots, F_r$ such that $|F_i \triangle F_j| = 2n$ for $1 \leq i < j \leq r$, then $|\mathcal{F}| < (2-\epsilon_r)^{4n}$.

To conclude this section let us consider the variant of the last problem in which $|F_i \triangle F_j|$ is replaced by $|F_i \cap F_j|$. This problem is solved by a fundamental result of design theory, namely *Fisher's inequality* (see Hall (1967)).

Theorem 8. *Let $\mathcal{F} = \{F_1, \ldots, F_m\}$ be a set system on X such that $|F_i \cap F_j| = k$ for all $i \neq j$. Then either $k = 0$ and $m \leq n+1$ or else $m \leq n$.*

Proof. The proof is very similar to the proof of Theorem 7. If $|F_i| = k$ for some i then $F_i \subset F_j$ for all j so $m - 1 \leq n - k$, implying $m \leq n + 1 - k$. Therefore we may assume that $a_i = |F_i| \geq k+1$ for all i.

Let $g_i = (g_{i1}, \ldots, g_{in}) \in \Re^n$ be the indicator function of F_i:

$$g_{ij} = \begin{cases} 1 & \text{if } j \in F_i \\ 0 & \text{otherwise.} \end{cases}$$

Then $|g_i|^2 = |F_i| = a_i \geq k+1$ and $(g_i, g_j) = |F_i \cap F_j| = k$ for all $i \neq j$.

We shall show that $g_1, \ldots, g_m \in \Re^n$ are linearly independent and so $m \leq n$. To this end, let

$$\sum_{i=1}^{m} \xi_i g_i = 0.$$

Then, as in the proof of Theorem 7,

$$0 = \sum_{i=1}^{m} \xi_i (g_i, g_j) = k \sum_{i=1}^{m} \xi_i + (a_j - k)\xi_j.$$

Hence, putting $\sum_{i=1}^{m} \xi_i = S$ we find that

$$\xi_j = \frac{k}{k - a_j} S. \tag{6}$$

Therefore

$$S = \sum_{j=1}^{m} \xi_j = S \sum_{j-1}^{m} \frac{k}{k - a_j}.$$

Since $\sum_{j-1}^{m} \frac{k}{k-a_j} < 0 < 1$, this implies that $S = 0$. Thus, by (6), $\xi_j = 0$ for every j, proving that $g_1, \ldots, g_m$ are linearly independent. ∎

Exercises

1. Check that if $\mathcal{F} \subset \mathcal{P}(X)$ is well-separated then so are $\mathcal{F}^c = \{X \setminus F : f \in \mathcal{F}\}$ and $\mathcal{F}\triangle\{x\} = \{F\triangle\{x\} : F \in X\}$, where $x \in X$.

2. Show that if $\mathcal{F} \subset \mathcal{P}([5])$ is well-separated then $|\mathcal{F}| \leq 4$. Show also that equality holds iff $\mathcal{F}$ is isomorphic to one of $\mathcal{F}_1$, $\mathcal{F}_2$, $\mathcal{F}_3$ and $\mathcal{F}_4$, where $\mathcal{F}_1 = \{12345, 12, 34, 5\}$, $\mathcal{F}_2 = \{1234, 125, 345, \emptyset\}$, $\mathcal{F}_3 = \{1235, 124, 45, 3\}$ and $\mathcal{F}_4 = \{135, 245, 14, 23\}$. Note also that $\mathcal{F}_2 = \mathcal{F}_1^c = \mathcal{F}\triangle\{5\}$, $\mathcal{F}_3 = \mathcal{F}_1\triangle\{4\}$ and $\mathcal{F}_4 = (\mathcal{F}_1\triangle\{1\})\triangle\{3\}$.

3. Let n be even and let $F_1, \ldots, F_m \subset X$ be such that $|F_i \triangle F_j| > n/2$ for all i, j, $1 \leq i < j \leq m$. Show that $m \leq n/2 + 1$. Show also that if $n \equiv 0 \bmod 4$ then $m \leq n/2$.

4. Let $\alpha < 1/2$ be fixed, and let $k = \lceil n/2 \rceil$, $A \in X^{(k)}$ and $\mathcal{B} = \{B \in X^{(k)} : d(A, B) < \alpha n\}$. Give an upper bound for $|\mathcal{B}|$. Use this upper bound to complete the proof of Theorem 6.

5. Let $\mathcal{F} = \{F_1, \ldots, F_m\}$ be a set system on X , and for $x \in X$ let $d(x)$, the *degree* of x, be the number of F_i's containing x. Suppose that $\emptyset$, $X \notin \mathcal{F}$, $d(x) < n$ for all $x \in X$, and if $x \notin F_i$ then $d(x) \leq |F_i|$. Prove that $m \leq n$. (Crawley and Dilworth (1973), see also Lovász (1979, p. 447))

(Suppose $m > n$. Check that if $x \notin F_i$ then

$$\frac{d(x)}{m - d(x)} < \frac{|F_i|}{n - |F_i|}.$$

Sum these inequalities over all pairs (x, F_i) such that $x \notin F_i$ and arrive at a contradiction.)

§11. HELLY FAMILIES

A set system $\mathcal{F} \subset \mathcal{P}(X)$ is said to be a *Helly family of order* k or an H_k*-family* if in every subfamily of $\mathcal{F}$ with empty intersection we can find a set consisting of at most k sets whose intersection is empty. Thus $\mathcal{F}$ is an H_k-family if for every subsystem $\mathcal{G} \subset \mathcal{F}$ with $\bigcap_{G \in \mathcal{G}} G = \emptyset$ we can find a subsystem $\mathcal{G}' \subset \mathcal{G}$ such that

$$\bigcap_{G \in \mathcal{G}'} G = \emptyset \qquad \text{and} \qquad |\mathcal{G}'| \le k.$$

An H_1-family is one in which the non-empty subsets have a common element, and an H_2-family is usually said to be a *Helly family.* Note that every family $\mathcal{F}$ is an H_k-family for all $k \ge |\mathcal{F}|$.

The terminology is inspired by Helly's classical theorem: if in a finite collection of convex sets in $\mathfrak{R}^{k-1}$ any k sets have a point in common then there is a point contained in all the sets. In other words, every (finite) family of convex sets in $\mathfrak{R}^{k-1}$ is an H_k-family (see Ex. 2).

Though Helly families are defined in terms of forbidden configurations, the definition is not too pleasant to work with since in general there are just too many of these forbidden configurations. Fortunately, it is rather pleasant to characterize those H_k-families which are known to be H_{k+1}-families. Suppose that $\mathcal{F}$ is an H_{k+1}-family. Then $\mathcal{F}$ is an H_k-family iff it does not contain sets $F_1, F_2, \ldots, F_{k+1}$ such that

$$\bigcap_{i=1}^{k+1} F_i = \emptyset \qquad \bigcap_{\substack{i=1 \\ i \ne j}} F_i \ne \emptyset,\ j = 1, 2, \ldots, k+1. \tag{1}$$

Indeed, if $\mathcal{G} \subset \mathcal{F}$ is such that $\bigcap_{G \in \mathcal{G}} = \emptyset$ then, as $\mathcal{F}$ is an H_{k+1}-family, there is a $\mathcal{G}_0 \subset \mathcal{G}$ such that $|\mathcal{G}_0| \le k+1$ and $\bigcap_{G \in \mathcal{G}_0} G = \emptyset$. Hence $\mathcal{F}$ is an H_k-family iff for every $\mathcal{G}_0 \subset \mathcal{F}$, $|\mathcal{G}_0| = k+1$, there is a $\mathcal{G}' \subset \mathcal{G}_0$ such that $|\mathcal{G}'| = k$ and $\bigcap_{G \in \mathcal{G}'} G = \emptyset$.

How large can an H_k family be? In view of (1), it is not surprising that the answer depends on the sizes of the sets in the family. The results below are due to Bollobás and Duchet (1979).

Before stating the theorems, let us note some easy conditions implying that a set system is an H_k-family.

(a) A subsystem of an H_k-family is itself an H_k-family.

(b) An H_{k-1}-family is also an H_k-family.

(c) If $\mathcal{F}$ is an H_k-family and $\mathcal{G} = \{F_1 \cap \ldots \cap F_t : F_i \in \mathcal{F}, t = 1, \ldots\}$ then $\mathcal{G}$ is an H_k-family. Every subfamily of $\mathcal{G}$ is again an H_k-family.

Of these observations, only (c) needs some justification. Suppose that $\bigcap_{i=1}^u G_i = \emptyset$ where $G_1, \ldots, G_u \in \mathcal{G}$. Then $\bigcap_{j=1}^v F_j = \emptyset$ for some $F_1, \ldots, F_v \in \mathcal{F}$, where each F_j contains at least one G_i, say G_{i_j}. But then some $l \le k$ of the F_j have empty intersection and therefore so do the corresponding G_{i_j}.

Theorem 1. *(i) If $\mathcal{F} \subset \bigcup_{s=0}^r X^{(s)}$ then $\mathcal{F}$ is an H_r-family iff it does not contain a $K_{r+1}^{(r)}$. In particular, $\mathcal{F}$ is an H_k-family for every $k \ge r+1$.*

(ii) If $k < r$ and $\mathcal{F} \subset X^{(r)}$ is an H_k-family then

$$|\mathcal{F}| \le \binom{n-1}{r-1}, \tag{2}$$

with equality iff $\mathcal{F} = X_x^{(r)}$ for some $x \in X$.

Proof. (i) Suppose $\mathcal{F} \subset \bigcup_{s=0}^r X^{(s)}$ is not an H_r-family. Then there is a $k \ge r$ such that $\mathcal{F}$ is an H_{k+1}-family but not an H_k-family. By (1), there are $F_1, F_2, \ldots, F_{k+1} \in \mathcal{F}$ such that $\bigcap_{i=1}^{k+1} F_i = \emptyset$ and $x_1, x_2, \ldots, x_{k+1} \in X$ such that $x_i \notin F_i$ and $x_j \in F_i$ for $j \ne i$. Consequently $G_i = \{x_j : j \ne i\} \subset F_i, i = 1, \ldots, k+1$. Since $|G_i| = k$ and $|F_i| \le r$, we have $k = r$ and $G_i = F_i$ for every i, $1 \le i \le r+1$. Thus $F_1, F_2, \ldots, F_{r+1}$ are the edges of a complete r-graph of order $r+1$.

The converse implication is immediate: the $r+1$ edges of $K_{r+1}^{(r)}$ have empty intersection but any r of the edges intersect, so the family formed by edges of $K_{r+1}^{(r)}$ is not an H_r-family.

(ii) Assume now that $k < r < n$ and $\mathcal{F} \subset X^{(r)}$ is an H_k-family.

Let us look at the "imprint" of $\mathcal{F}$ on $X^{(r-1)}$. Set

$$\mathcal{A} = \{A \in X^{(r-1)} : A = F_1 \cap F_2 \text{ for some } F_1, F_2 \in \mathcal{F}\},$$

By observation (c), we know that $\mathcal{A}$ is an H_k-family. Hence, by part (i) of our theorem, $\mathcal{A}$ does not contain a $K_r^{(r-1)}$. In particular, every $F \in \mathcal{F}$ has a subset $A_F \in X^{(r-1)}$ which is contained in no other $F \in \mathcal{F}$. As noted by Tuza (1984), Theorem 9.2 of Bollobás (1965) enables us to complete the proof. First let us note another trivial reformulation of Theorem 9.2.

Given a set system $\mathcal{F}$ and a set $F \in \mathcal{F}$, call a set $A_F \subset F'$ an *own subset of* F (for $\mathcal{F}$) if $A_F \not\subset F'$ for $F' \in \mathcal{F}$, $F' \neq F$. The condition in Theorem 9.2 says precisely that the system $\mathcal{F} = \{F_i = A_i \cup B_i : i \in I\}$ is such that A_i is an own subset of F_i and the conclusion is that

$$\sum_{F \in \mathcal{F}} \binom{n - |F| + |A_F|}{|A_F|}^{-1} \leq 1 \qquad (3)$$

with equality iff $\mathcal{F} = \{F \in X^{(s)} : F \supset B\}$ for a set $B \subset X$ and an integer s, $|B| \leq s \leq n$, and $A_F = F \setminus B$ for $F \in \mathcal{F}$.

As in our case $\mathcal{F} \subset X^{(r)}$ and $|A_F| = r - 1$ for every $F \in \mathcal{F}$, by (3) we have

$$|\mathcal{F}| \binom{n - r + r - 1}{r - 1}^{-1} \leq 1$$

so

$$|\mathcal{F}| \leq \binom{n-1}{r-1}$$

as required. Equality holds iff we have equality in (3) and so $\mathcal{F} = X_x^{(r)}$ for some $x \in X$. ∎

What does Theorem 1 tell us about the maximal size of an H_k-family contained in $X^{(r)}$? If $k < r$ then the maximum is $\binom{n-1}{r-1}$ and if $k > r$ then it is $\binom{n}{r}$. However, for $k = r$ it is precisely the maximal size of an r-graph without a $K_{r+1}^{(r)}$, so we cannot expect to determine it precisely. It is a little surprising that it is easy to determine the maximal size of an H_k-family contained in $X^{(k)} \cup X^{(k+1)}$.

Theorem 2. *Let $\mathcal{F} \subset X^{(k)} \cup X^{(k+1)}$ be an H_k-family. Then*

$$|\mathcal{F}| \leq \binom{n}{k}.$$

The inequality is best possible.

Proof. Similarly to the proof of Theorem 1, set

$$\begin{aligned} \mathcal{A} &= \{A \in X^{(k)} : A = A_1 \cap A_2 \text{ for some } A_1, A_2 \in \mathcal{F}_{k+1}\}, \\ \mathcal{A}' &= \mathcal{A} \cup \mathcal{F}_k, \\ \mathcal{B}' &= X^{(k)} \setminus \mathcal{A}'. \end{aligned}$$

By our observation (c), the system $\mathcal{A}'$ is an H_k-family. Therefore, by Theorem 1(i), it does not contain a complete k-graph of order $k+1$, and so every member of $\mathcal{F}_{k+1}$ contains a member of $\mathcal{B}'$ implying

$$|\mathcal{F}_{k+1}| \leq |\mathcal{B}'|.$$

Consequently

$$|\mathcal{F}_k| + |\mathcal{F}_{k+1}| \leq |\mathcal{A}'| + |\mathcal{B}'| \leq \binom{n}{k}.$$

To see that the inequality is best possible, pick an element $x \in X$ and note that $\mathcal{F} = X_x^{(k)} \cup X_x^{(k+1)}$ is an H_1-family and

$$|\mathcal{F}| = \binom{n-1}{k-1} + \binom{n-1}{k} = \binom{n}{k}. \qquad \blacksquare$$

Corollary 3. *Let* $1 \leq k < r \leq n$. *If* $\mathcal{F} \subset X^{(\leq r)} = \bigcup_{s=0}^{r} X^{(s)}$ *is an* H_k*-family then*

$$|\mathcal{F}| \leq \sum_{i=0}^{k-1} \binom{n}{i} + \sum_{j=k}^{r} \binom{n-1}{j-1}. \tag{4}$$

The inequality is best possible.

Proof. Set $\mathcal{F}_i = \mathcal{F} \cap X^{(i)}$. By Theorem 1 we have

$$\left| \sum_{s=k+2}^{r} \mathcal{F}_s \right| \leq \sum_{s=k+2}^{r} \binom{n-1}{s-1}$$

and, by Theorem 2,

$$|\mathcal{F}_k \cup \mathcal{F}_{k+1}| \leq \binom{n}{k} = \binom{n-1}{k-1} + \binom{n-1}{k}.$$

Since $|\mathcal{F}_i| \leq \binom{n}{i}$ for every i, our inequality holds.

To see that the inequality is best possible, pick an $x \in X$ and set $\mathcal{F} = \bigcup_{i=0}^{k-1} X^{(i)} \cup \bigcup_{j=k}^{r} X_x^{(j)}$. Then $\mathcal{F}$ is an H_k-family since every $F \in \mathcal{F}$ that has at least k elements contains the base point x. ∎

In fact, the set system above is the unique set system for which we have equality in (4) (see Ex. 4).

Exercises

1. Let $A = \{a_1, a_2, \ldots, a_{k+1}\} \subset \Re^{(k-1)}$. Show that there is a point $a \in \Re^{(k-1)}$ which is a convex linear combination of any k of the points a_i, i.e. which is such that for every j we have $a = \sum_{i=1}^{k+1} t_i a_i$ for some $t_i \geq 0$ satisfying $\sum_{i=1}^{k+1} t_i = 1$ and $t_j = 0$.
(Apply induction on k. Show that for some $1 \leq i < j \leq k+1$ there is a point $a_{ij} = \lambda a_i + \mu a_j$ with $\lambda, \mu \geq 0$, $\lambda + \mu = 1$, in the affine plane determined by $A' = (A \setminus \{a_i, a_j\})$.)

2. Deduce Helly's theorem from Exercise 1: if a finite family of convex sets in $\Re^{(k-1)}$ is such that any k of the C_i's have a point in common then some point is contained in all the sets.

3. Is Helly's theorem true for infinite families of convex sets?

4. Check that $\mathcal{F} = X^{(\leq k-1)} \cup \bigcup_{j=k}^{r} X_x^{(j)}$ is the only extremal set system for Corollary 3. (Bollobás and Duchet (1983))

5. Prove that if $|X| = n \geq 5$, $r \geq 3$ and $\mathcal{F} \subset X^{(\leq r)}$ is a Sperner system which is also an H_2-family then

$$|\mathcal{F}| \leq \binom{n-1}{r-1} \qquad \text{if } r \leq n/2,$$

$$|\mathcal{F}| \leq \binom{n-1}{\lfloor n/2 \rfloor} \qquad \text{if } r > n/2.$$

(Bollobás and Duchet (1983))

6. Check that the proof of Theorem 1 (ii) gives the following strengthening of inequality (2): if $\mathcal{F} \subset \bigcup_{s=k+1}^{n} X^{(s)}$ is a Sperner system which is also an H_k-family then

$$\sum_{F \in \mathcal{F}} \binom{n-1}{|F|-1}^{-1} \leq 1.$$

(Tuza (1984))

7. Deduce from Turán's theorem for graphs (see §8) that the maximal size of a Sperner H_2-family $\mathcal{F} \subset X^{(\leq 2)}$ is precisely $\lfloor n^2/4 \rfloor$.

§12. HYPERGRAPHS WITH A GIVEN NUMBER OF DISJOINT EDGES

In Theorem 7.1 we determined the maximal size of an r-graph without two disjoint edges. In this section we shall examine a more general question: at most how many edges are there in an r-graph of order n if the graph contains at most k pairwise disjoint edges? If n is small, namely if $n < (k+1)r$, then no r-graph contains $k+1$ pairwise disjoint edges, so in this case the maximum is $\binom{n}{r}$. However, for $n \geq (k+1)r$ the problem becomes very interesting indeed. For given values of r and k, a complete solution was given by Erdős (1965) for every large enough n. Not unexpectedly, the extremal hypergraphs are of the form $X^{(r)}(x_1, x_2, \ldots, x_k) = \{A \in X^{(r)} : \{x_1, x_2, \ldots, x_k\} \cap A \neq \emptyset\}$. We express this by saying that the extremal graphs are formed by all r-sets *fixed by k given vertices.* Of course, it is readily seen that $X^{(r)}(x_1, \ldots, x_k)$ does not contain $k+1$ pairwise disjoint edges, so the problem is to prove that this is the (essentially) unique best possible construction.

In fact, instead of presenting this result of Erdős, we shall prove a theorem of Hajnal and Rothschild (1973) which extends not only this result of Erdős, but Theorem 7.3 as well.

We call $\mathcal{A} \subset \mathcal{P}(X)$ a *(k,l)-intersecting family* if whenever $A_1, A_2, \ldots, A_{k+1}$ are $k+1$ distinct members of $\mathcal{A}$, we have $|A_i \cap A_j| \geq l$ for some i and j, $1 \leq i < j \leq k+1$. In other words, $\mathcal{A}$ is a (k,l)-intersecting family if it contains at most k sets such that the intersection of any two of them has fewer than l elements. We shall always take $k \geq 1$ and $l \geq 1$ because every family $\mathcal{A}$ is $(k,0)$-intersecting for $k \geq 1$ and a family $\mathcal{A}$ is $(0,l)$-intersecting iff $\mathcal{A} = \emptyset$. Clearly every (k,l)-intersecting family is $(k+1,l)$-intersecting and $(k,l+1)$-intersecting .

In this section we shall study (k,l)-intersecting r-graphs. Note that a $(1,l)$-intersecting r-graph is precisely an l-intersecting r-graph and, at the other extreme, $\mathcal{A}$ is a $(k,1)$-intersecting r-graph iff it does not contain

$k+1$ pairwise disjoint edges.

What is a fool-proof way of constructing a (k,l)-intersecting r-graph? Take k not necessarily disjoint l-subsets: $L_1, L_2, \ldots, L_k$, and make sure that every edge of your graph $(X, \mathcal{A})$ contains at least one L_i: let $\mathcal{A} = X^{(r)}(L_1, \ldots, L_k) = \{A \in X^{(r)} : A \supset L_i \text{ for some } i\}$. We shall express this by saying that $\mathcal{A}$ is *fixed by the l-subsets* $L_1, L_2, \ldots, L_k$. Then $\mathcal{A}$ is indeed a (k,l)-intersecting family, for if $A_1, A_2, \ldots, A_{k+1} \in \mathcal{A}$ then each A_i contains some L_{j_i}, $1 \le j_i \le k$, so at least one L_j is contained in at least two A_i's, and the intersection of these two A_i's has at least l elements.

We shall show that, for given r, k and l, $1 \le l < r$, $l \ge 1$, and every sufficiently large n, if $\mathcal{A} \subset X^{(r)}$ is (k,l)-intersecting then

$$|\mathcal{A}| \le |X^{(r)}(L_1, L_2, \ldots, L_k)| \tag{1}$$

for some l-subsets $L_1, L_2, \ldots, L_k$. Then it will be an easy matter to show that, under certain conditions, the L_i's may be chosen to be disjoint .

Note that if $k = 1$ then inequality (1) has already been proved. Indeed, if $k = 1$ then by Theorem 7.3, $|\mathcal{A}| \le |X^{(r)}(L_1)|$, where L_1 is an l-set, provided n is sufficiently large.

How can we tell whether a system $\mathcal{A} \subset X^{(r)}$ is (k,l)-intersecting or not? A rather natural way is to take k sets $A_1, A_2, \ldots, A_k \in \mathcal{A}$ such that no two A_i's share an l-set and try to add to it another $A \in \mathcal{A}$ with this property. The system $\mathcal{A}$ is (k,l)-intersecting if we are never successful in this attempt.

Let us formalize this condition. Define

$$\mathcal{B} = \mathcal{B}(\mathcal{A}, k, l) = \Big\{ \bigcup_{i=1}^{k} A_i^{(l)} : A_i \in \mathcal{A}, A_i^{(l)} \cap A_j^{(l)} = \emptyset \text{ for } i \ne j \Big\}. \tag{2}$$

Note that every $B \in \mathcal{B}$ is a subset of $Y = X^{(l)}$ and, in fact, $|B| = k\binom{r}{l}$. Thus $\mathcal{B} \subset Y^{(b)}$, where $b = k\binom{r}{l}$. The r-graph $(X, \mathcal{A})$ is $(k-1,l)$-intersecting iff $\mathcal{B} = \emptyset$ and it is (k,l)-intersecting iff

$$A^{(l)} \cap B \ne \emptyset \quad \text{for all } A \in \mathcal{A} \text{ and } B \in \mathcal{B}. \tag{3}$$

If $\mathcal{A}$ is fixed by the l-subsets $L_1, L_2, \ldots, L_k$ then $L_i \in B$ for all i, $1 \le i \le k$, and all $B \in \mathcal{B}$. (The elements of a set $B \in \mathcal{B}$ are l-sets!) Hence in this case $|\bigcap_{B \in \mathcal{B}} B| \ge k$. We shall show now that if $\mathcal{A}$ is very far from being fixed by k l-sets in the sense that $\bigcap_{B \in \mathcal{B}} B = \emptyset$ then $|\mathcal{A}|$ is considerably smaller than $|X^{(r)}(L_1,\ldots,L_k)|$, provided n is sufficiently large, where $L_1,\ldots,L_k$, are disjoint l-subsets of X.

Lemma 1. *Let $1 \le l < r < n$ and $k \ge 1$. Suppose $(X,\mathcal{A})$ is a (k,l)-intersecting r-graph but it is not $(k-1,l)$-intersecting. If $\bigcap_{B\in\mathcal{B}} B = \emptyset$ then*

$$|\mathcal{A}| \le c_k n^{r-l-1}$$

where $c_k = c(r,k,l) = \binom{u}{l+1}\Big/(r-l-1)!$ and $u = k^2 r\binom{r}{l} + kr$.

Proof. As $\mathcal{A}$ is not $(k-1,l)$-intersecting, $\mathcal{B} \ne \emptyset$. Since $\bigcap_{B\in\mathcal{B}} B = \emptyset$ and $|B| = b = k\binom{r}{l}$ for every $B \in \mathcal{B}$, we can find a set $\mathcal{B}' \subset \mathcal{B}$ such that $|\mathcal{B}'| \le b+1$ and

$$\bigcap_{B\in\mathcal{B}'} B = \emptyset. \tag{4}$$

Indeed, pick $B_0 \in \mathcal{B}$, say $B_0 = \{L_1, L_2, \ldots, L_b\}$, and for each i, $1 \le i \le b$, pick a set $B_i \in \mathcal{B}$ not containing L_i. Then $\mathcal{B}' = \{B_i : 0 \le i \le b\}$ will do. (Though this seems to show that $|\mathcal{B}'| = b+1$, we have only $|\mathcal{B}'| \le b+1$ since $B_i = B_j$ may hold for $i \ne j$.) In fact, $\mathcal{B}'$ exists precisely because a family of b-subsets has the Helly property of order $b+1$, as claimed by the easy part of Lemma 11.1(i).

For $B \in \mathcal{B}$ pick a representation $B = \bigcup_{i=1}^k A_i^{(l)}$, $A_i^{(l)} \cap A_j^{(l)} = \emptyset$ for $1 \le i < j \le k$, and set $U(B) = \bigcup_{i=1}^k A_i$. Clearly $|U(B)| \le kr$. Let

$$U = \bigcup_{B\in\mathcal{B}'} U(B).$$

Then

$$|U| \le kr|\mathcal{B}'| \le kr(b+1) = u.$$

We claim that every $A \in \mathcal{A}$ satisfies

$$|A^{(l)} \cap \bigcup_{B\in\mathcal{B}'} B| \ge 2. \tag{5}$$

Indeed, by (3),

$$A^{(l)} \cap B \ne \emptyset \qquad \text{for all } B \in \mathcal{B},$$

so if (5) fails then $A^{(l)} \cap \bigcup_{B\in\mathcal{B}'} B = \{L\}$ and $L \in B$ for all $B \in \mathcal{B}'$. (The formula $L \in B$ seems wrong; to see that it can be correct, remember that L is an l-set and B is a set of l-sets.) But then $L \in \bigcap_{B\in\mathcal{B}'} B$, contradicting (4).

Now (5) implies that at least two l-subsets of A are contained in U. As the union of two distinct l-sets has at least $l+1$ elements, we find that

$$|A \cap U| \geq l+1 \tag{6}$$

for all $A \in \mathcal{A}$.

Relation (6) gives the required bound on $|\mathcal{A}|$, the total number of sets $A \in X^{(r)}$ satisfying (6) is at most

$$\binom{|U|}{l+1}\binom{n-l-1}{r-l-1} \leq c_1 n^{r-l-1}$$

where the first factor is the number of ways we can select $l+1$ elements of U and the second is the number of ways $r-l-1$ more elements can be selected from X. ■

Lemma 1 enables us to deduce an assertion which is very close to our main aim.

Lemma 2. *Suppose $1 \leq l < r < n$, $k \geq 1$, and $(X, \mathcal{A})$ is a (k,l)-intersecting r-graph not fixed by k l-sets. Then*

$$|\mathcal{A}| \leq (k-1)\binom{n-l}{r-l} + c_k n^{r-l-1} \tag{7}$$

where $c_k = c(r,k,l)$ is as in Lemma 1.

Proof. Let us apply induction on k. Consider first $k=1$. If $\mathcal{A} = \emptyset$ the assertion is trivial. Hence we may assume that $\mathcal{A} \neq \emptyset$, i.e. $\mathcal{A}$ is not $(0,1)$-intersecting. Since $\bigcap_{A \in \mathcal{A}} A^{(l)} = \emptyset$, by Lemma 1 we have

$$|\mathcal{A}| \leq c_k n^{r-l-1},$$

as required.

Suppose then that $k \geq 2$ and the assertion holds for smaller values of k. What can we assume about $\mathcal{A}$? Firstly, we may assume that $\mathcal{A}$ is not $(k-1,l)$-intersecting. Indeed, otherwise either $\mathcal{A}$ is fixed by $k-1$ l-sets and so

$$|\mathcal{A}| \leq (k-1)\binom{n-l}{r-l}$$

since the number of r-sets containing a given l-set is $\binom{n-l}{r-l}$, or else $\mathcal{A}$ is not fixed by $k-1$ l-sets and then, by the induction hypothesis,

$$|\mathcal{A}| \leq (k-2)\binom{n-l}{r-l} + c_{k-1} n^{r-l-1}$$
$$\leq (k-2)\binom{n-l}{r-l} + c_k n^{r-l-1}.$$

Secondly, we may assume that $L_1 \in \bigcap_{B \in \mathcal{B}} B$ for some l-set L_1, where $\mathcal{B} = \mathcal{B}(\mathcal{A}, k, l)$ is given by (2), since otherwise by Lemma 1 we have

$$|\mathcal{A}| \leq c_k n^{r-l-1}.$$

Let us use this l-set L_1 to partition $\mathcal{A}$:

$$\mathcal{A}' = \{A \in \mathcal{A} : L_1 \subset A\},$$
$$\mathcal{A}'' = \{A \in \mathcal{A} : L_1 \not\subset A\} = \mathcal{A} \setminus \mathcal{A}'.$$

We shall estimate $|\mathcal{A}'|$ and $|\mathcal{A}''|$ separately. Clearly

$$|\mathcal{A}'| \leq \binom{n-l}{r-l}. \tag{8}$$

In order to estimate $|\mathcal{A}''|$, note that $\mathcal{A}'' \subset X^{(r)}$ is a $(k-1, l)$-intersecting family, for otherwise there exist $A_1, A_2, \ldots, A_k \in \mathcal{A}''$ such that the sets $A_i^{(l)}, i = 1, 2, \ldots, k$ are disjoint. But then $B_0 = \bigcup_{i=1}^k A_i^{(l)} \in \mathcal{B}$ and $L_1 \notin B_0$, contradicting our assumption that L_1 belongs to every member of $\mathcal{B}$.

Is $\mathcal{A}''$ fixed by $k-1$ l-sets? Certainly not, for if $\mathcal{A}''$ is fixed by the l-sets $L_2, L_3, \ldots, L_k$ then $\mathcal{A}$ is fixed by the l-sets $L_1, L_2, \ldots, L_k$. Therefore, by the induction hypothesis,

$$|\mathcal{A}''| \leq (k-2)\binom{n-l}{r-l} + c_{k-1} n^{r-l-1} \tag{9}$$

As $|\mathcal{A}| = |\mathcal{A}'| + |\mathcal{A}''|$, inequalities (8) and (9) imply (7). ∎

Theorem 3. *Suppose $1 \leq l < r < n$, $k \geq 1$, and $\mathcal{A} \subset X^{(r)}$ is a (k, l)-intersecting family. Then there is a number $n(r, k, l)$ such that if $n \geq n(r, k, l)$ then*

$$|\mathcal{A}| \leq \sum_{j=1}^{k} (-1)^{j+1} \binom{k}{j} \binom{n - jl}{r - jl}.$$

Furthermore, equality holds iff $\mathcal{A} = X^{(r)}(L_1, L_2, \ldots, L_k)$, where the L_i's are l-subsets of X such that if $L_i \cap L_j \neq \emptyset$ and $i \neq j$ then $|L_i \cup L_j| \geq r+1$. In particular, if $r \geq 2l-1$ then equality holds iff $\mathcal{A} = X^{(r)}(L_1, L_2, \ldots, L_k)$ for some disjoint l-subsets $L_1, \ldots, L_k$ of X.

Proof. Let $n \geq kl$ and let $L_1, L_2, \ldots, L_k$ be disjoint l-sets in X. Set $\mathcal{A}_0 = X^{(r)}(L_1, \ldots, L_k)$. As there are precisely $\binom{n-jl}{r-jl}$ elements of $X^{(r)}$ that contain j given disjoint l-subsets of X, by the inclusion-exclusion principle we have

$$
\begin{aligned}
|\mathcal{A}_0| &= \sum_{j=1}^{k} (-1)^{j+1} \binom{k}{j} \binom{n-jl}{r-jl} \\
&\geq k \binom{n-l}{r-l} - dn^{r-l-1},
\end{aligned}
$$

where $d = d(r,k,l)$. This shows that if $|\mathcal{A}| \geq |\mathcal{A}_0|$ then, by Lemma 2, $\mathcal{A}$ is fixed by some k l-sets, provided n is sufficiently large.

All we have to do then is to prove that if n is sufficiently large and $L_1, \ldots, L_k$ are l-subsets of X then

$$|X^{(r)}(L_1, \ldots, L_k)| \leq |\mathcal{A}_0| \tag{10}$$

and characterize the k-tuples $(L_1, \ldots, L_k)$ for which equality holds.

We shall prove (10) by changing a non-disjoint system $L_1, L_2, \ldots, L_k$ step-by-step into a disjoint system in such a way that in no step does $|X^{(r)}(L_1, \ldots, L_k)|$ decrease. Suppose that $n \geq kl$, $L_1, \ldots, L_k \subset X$ are l-sets and, say, $a \in L_1 \cap L_2$. Pick an element $b \in X \setminus \bigcup_{i=1}^{k} L_i$; since $|\bigcup_{i=1}^{k} L_i| \leq kl - 1 < n$, this can be done. Set $L_1' = (L_1 \setminus \{a\}) \cup \{b\}$, $\mathcal{A} = X^{(r)}(L_1, L_2, \ldots, L_k)$ and $\mathcal{A}' = X^{(r)}(L_1', L_2, \ldots, L_k)$. We claim that

$$|\mathcal{A}| \leq |\mathcal{A}'|. \tag{11}$$

To see (11), note that for each set A we lose from $\mathcal{A}$ we gain a new set A' in $\mathcal{A}'$. Indeed,

$$\mathcal{A} \setminus \mathcal{A}' = \{A \in X^{(r)} : A \supset L_1, A \not\supset L_i \quad \text{for } i \geq 2, b \notin A\}$$

and

$$\mathcal{A}' \setminus \mathcal{A} = \{A \in X^{(r)} : A \supset L_1', A \not\supset L_i \quad \text{for } i \geq 2, a \notin A\}.$$

Hence to each $A \in \mathcal{A} \setminus \mathcal{A}'$ there corresponds the set $A' = (A \setminus \{a\}) \cup \{b\} \in \mathcal{A}' \setminus \mathcal{A}$. Since this map is one-to-one,

$$|\mathcal{A} \setminus \mathcal{A}'| \leq |\mathcal{A}' \setminus \mathcal{A}|, \tag{12}$$

proving (11).

Inequality (10) is an immediate consequence of (11) since if $|\bigcup_{i=1}^{k} L_i| = kl - m$ then m steps of the form $(L_1, L_2, \ldots, L_k) \rightarrow (L_1', L_2, \ldots, L_k)$ result in a system of disjoint L_i's. When do we have equality in (10)? If no r-set can contain two L_i's then this is certainly the case, for then

$$|X^{(r)}(L_1, L_2, \ldots, L_k)| = k\binom{n-l}{r-l}.$$

Furthermore, trivially, we have equality if the L_i's are disjoint.

We claim that if $n \geq r + k - 1$ then in every other case we have strict inequality in (10). Suppose then that $L_1, L_2, \ldots, L_k$ are such that $L_1 \cap L_2 \neq \emptyset$ and $|L_1 \cup L_2| \leq r$. As in our steps we do not decrease the cardinality of $\mathcal{A} = X^{(r)}(L_1, L_2, \ldots, L_r)$, we may assume that $L_3, L_4, \ldots, L_k$ have already been made disjoint: $L_i \cap L_j \neq \emptyset$ with $1 \leq i < j \leq k$ iff $i = 1$ and $j = 2$. Let a, b and $\mathcal{A}'$ be as before. Consider a set $B \in X^{(r)}$ such that $B \supset \big((L_1 \cup L_2) \setminus \{a\}\big) \cup \{b\}$ and $B \not\supset L_i$ for $i \geq 3$. There is such a B since X is large enough: $n \geq r + k - 1$. Now this set B belongs to $\mathcal{A}' \setminus \mathcal{A}$ but it is not of the form $B = A'$ for some $A \in \mathcal{A} \setminus \mathcal{A}'$ because $A = (B \setminus \{b\}) \cup \{a\} \in \mathcal{A}'$. Therefore we have strict inequality in (12) and so (10) is strict, as claimed. ∎

In §19 we shall prove the analogue of Theorem 3 for set systems.

Exercises

1. For $\mathcal{F} \subset \mathcal{P}(X)$ let $p(\mathcal{F})$ be the number of disjoint pairs of sets in $\mathcal{F}$. Let $\tilde{S}_{ij}$ be the compression operator used in §5. Prove that

$$p(\partial\mathcal{F}) = p(\partial_l\mathcal{F}) \leq p(\mathcal{F}).$$

2. Prove the following theorem of Katona (1964). If $\mathcal{F} \subset X^{(r)}$ is an intersecting r-graph then

$$|\partial\mathcal{F}| \geq |\mathcal{F}|.$$

3. (Harder) Let $\mathcal{F}$ be an intersecting r-graph, $r \geq 2$. Prove that there is a set M of size at most $(2r-1)\binom{2r-3}{r-1}$ such that $F_1 \cap M \cap F_2 \neq \emptyset$ for all $F_1, F_2 \in \mathcal{F}$.

(This is a considerable sharpening by Lovász (1979, p. 456) of a result of Całczyńska-Karlowicz (1964). To prove it, following Lovász, assume that

$\mathcal{F} \subset X^{(\le r)}$ is an intersecting family whose ground set is minimal, i.e. for every $x \in X$ there are distinct sets $F_1, F_2 \in \mathcal{F}$ such that $F_1 \cap F_2 = \{x\}$. We have to show that $n \le (2r-1)\binom{2r-3}{r-1}$.

If $|F| = 1$ for some $F \in \mathcal{F}$ then $n = 1$ so assume that $\mathcal{F} \subset X^{(\ge 2)}$. Consider all $n!$ orderings of X. Check that for each ordering $x_1, x_2, \ldots, x_n$ there is at most one i such that both $\{x_1, \ldots, x_i\}$ and $\{x_i, \ldots, x_n\}$ contain sets of $\mathcal{F}$, say $F_1 \subset \{x_1, \ldots, x_i\}$, $F_2 \subset \{x_i, \ldots, x_n\}$ and $F_1 \cap F_2 = \{x_i\}$. On the other hand, if x, F_1 and F_2 intersect in x, $|F_1| = s$ and $|F_2| = t$, then x plays the role of x_i in at least

$$\begin{aligned} 2(s-1)!(t-1)!\binom{n}{s+t-1}(n-s-t+1)! &= 2n!\frac{(s-1)!(t-1)!}{(s+t-1)!} \\ &\ge 2\frac{n!}{r}\binom{2r-1}{r}^{-1} \end{aligned}$$

orderings, the factor 2 being due to the fact that either F_1 precedes F_2 or else F_2 precedes F_1.)

4. (Ex. 3 continued) Can one have equality for $r = 2$? And for $r = 3$?

5. (Ex. 3 continued) Show that the same assertion holds if $\mathcal{F}$ is an infinite intersecting r-graph.

(Let $m(\mathcal{F}')$ be the minimal size of a set M that will do for a finite subgraph $\mathcal{F}' \subset \mathcal{F}$. Pick a finite subgraph $\mathcal{F}_0 \subset \mathcal{F}$ such that $m(\mathcal{F}_0) = \max\{m(\mathcal{F})' : \mathcal{F}' \subset \mathcal{F},\ \mathcal{F}' \text{ is finite}\}$. Show that some subset of the union $\bigcup\{F : F \in \mathcal{F}_0\}$ will do.)

§13. INTERSECTING FAMILIES

We remarked in §7 that if any two sets in a family $\mathcal{F} \subset \mathcal{P}(X)$ intersect then, rather trivially, $|\mathcal{F}| \le 2^{n-1}$. This leads us to considering intersecting hypergraphs, i.e. r-graphs without two disjoint edges, and to the first Erdős, Ko, Rado theorem (Theorem 7.1). As an extension of the concept of an intersecting family, we introduced l-intersecting families: set systems in which any two sets share at least l elements. The second Erdős, Ko, Rado theorem (Theorem 7.3) gave an essentially best possible bound on the size of an l-intersecting r-graph. Further extensions of these theorems were proved in §12.

In this section we shall return to set systems without any restrictions on the sizes of the sets. First we shall present an extension of the first Erdős, Ko, Rado theorem to Sperner families with appropriate weights and then we shall examine the maximal size of an l-intersecting set system.

The first Erdős, Ko, Rado theorem (Theorem 7.1) concerns intersecting hypergraphs. Now a hypergraph $\mathcal{A} \subset X^{(r)}$ is a Sperner family (or an antichain) so it is rather natural to seek an extension to Sperner families along the lines of Theorems 3.2 and 9.2. Such an extension was obtained by Bollobás (1973), who made use of Katona's proof of Theorem 7.1. As in that proof, we arrange the elements of $X = [n]$ in a cyclic order and consider only intervals, i.e. sets of consecutive elements. To make this more precise, let $T = \mathbf{Z}_n = \{0, 1, \dots, n-1\}$ be the group of integers modulo n and call a set $J \subset T$ an *interval* if it is of the form $J = \{k, k+1, \dots, k+l\}$ for some $k, l \in T$. Write $\mathcal{J}(T)$ for the system of all intervals and $\mathcal{J}^{(m)}(T)$ for the set of all intervals of size m.

Lemma 1. *Let $\mathcal{B} \subset \mathcal{J}(T)$ be an intersecting Sperner system with* $\max\{|B| : B \in \mathcal{B}\} \le n/2$ *and* $\min\{|B| : B \in \mathcal{B}\} = m$. *Then*

$$|\mathcal{B}| \le m.$$

Furthermore, if $m < n/2$ then $\mathcal{B} \subset \mathcal{J}(T)$ and $|\mathcal{B}| = m$ iff

$$\mathcal{B} = \mathcal{J}_t^{(m)} = \{B : B \in \mathcal{J}^{(m)}(T), B \ni t\} \quad \text{for some } t \in T.$$

Proof. Let $B_0 \in \mathcal{B} \cap \mathcal{J}^{(m)}(T)$, $B_0 = \{k, k+1, \ldots, k+m-1\}$, and for $B \in \mathcal{B}$ let $b = \beta(B)$ be the unique element of $M = \{k+1, k+2, \ldots, k+m-1\}$ for which $|B \cap \{b-1, b\}| = 1$. Then $\beta : \mathcal{B} - \{B_0\} \to M$ is a one-to-one map since if $\beta(B_1) \cap \{b-1, b\} = \{b-1\}$ and $\beta(B_2) \cap \{b-1, b\} = \{b\}$ then $B_1 \cap B_2 = \emptyset$ since $|B_1| + |B_2| \leq n$, and if $\beta(B_1) \cap \{b-1, b\} = \beta(B_2) \cap \{b-1, b\}$ then $B_1 \subset B_2$ or $B_2 \subset B_1$.

As β is one-to-one, $|\mathcal{B}| \leq |M| + 1 = m$. The second assertion is clear. ∎

Theorem 2. *Let $\mathcal{A} \subset \mathcal{P}(X)$ be an intersecting Sperner system with $|A| \leq n/2$ for all $A \in \mathcal{A}$. Then*

$$\sum_{A \in \mathcal{A}} \binom{n-1}{|A|-1}^{-1} \leq 1. \tag{1}$$

Equality holds iff $\mathcal{A} = X_x^{(r)}$ for some $r < n/2$ or n is even and $\mathcal{A}$ contains precisely one of each pair of complementary $\frac{n}{2}$-sets.

Proof. Denote by N the number of one-to-one maps from X to T and let $n(A)$ be the number of those one-to-one maps $\psi : X \to T$ for which $\psi(A)$ is an interval. Then, by Lemma 1,

$$\sum_{A \in \mathcal{A}} \frac{n(A)}{|A|} \leq N. \tag{2}$$

Clearly $N = n!$ and $n(A) = a!\,(n-1)!\,n$ if $a = |A|$ since there are n intervals of length a. Thus, $n(A)/(|A|N) = \frac{(a-1)!\,(n-a)!}{n!}$ and so (1) follows from (2).

It is clear that if $\mathcal{A} = X_x^{(r)}$ for some $r \leq n/2$ then equality holds in (1). Conversely, if equality holds in (1) then, by Lemma 1, for every one-to-one map $\psi : X \to T$ we must have $\tilde{\psi}(\mathcal{F}) = \{\psi(A) : A \in \mathcal{F}$ and $\psi(A)$ is an interval$\} = \mathcal{J}_t^{(m)}(T)$ for some $m = m(\psi)$ and $t = t(\psi)$, or else $\tilde{\psi}(\mathcal{F}) = \mathcal{J}^{(n/2)}(T)$. As in the second proof of Theorem 7.1, it is easily checked that either $\mathcal{A} \subset X^{(n/2)}$ or else neither $m(\psi)$ nor $T(\psi)$ depend on ψ and $\mathcal{A} = X_x^{(r)}$, as claimed. ∎

The reader may have noticed that we deduced (1) from Lemma 1 by repeating the proof of Theorem 7.3 in this special case. In particular, (1) is immediate from Lemma 1 and Theorem 7.3.

Note that Theorem 2 implies the following slight extension of Theorem 7.1: if $r < n/2$ and $\mathcal{A} \subset X^{(\leq r)}$ is an intersecting Sperner system then

$$|\mathcal{A}| \leq \binom{n-1}{r-1},$$

with equality iff $\mathcal{A} = X_x^{(r)}$.

Clearly Theorem 2 has the following consequence which is hardly more than a reformulation of the result.

Theorem 3. *Let $\mathcal{F}$ be a Sperner system of subsets of X such that if $A \in \mathcal{F}$ then $X \setminus A \in \mathcal{F}$. Then*

$$\sum_{A \in \mathcal{F}} w(A) \leq 2$$

where for $|A| = r$ the weight $w(A)$ of A is defined by

$$w(A) = \max\left\{ \binom{n-1}{r-1}^{-1}, \binom{n-1}{r}^{-1} \right\}.$$

Proof. Let $\mathcal{A} \subset \mathcal{F} \cap X^{(\leq n/2)}$ be such that for $\{A, X \setminus A\} \subset \mathcal{F}$ the system $\mathcal{A}$ contains precisely one of A and $X \setminus A$. Then, by Theorem 2,

$$\sum_{A \in \mathcal{F}} w(A) = 2 \sum_{A \in \mathcal{A}} \binom{n-1}{|A|-1}^{-1} \leq 2. \qquad \blacksquare$$

Let us turn to the second problem to be discussed in this section: let us examine the maximal size of an l-intersecting set system. What set systems are easily seen to be l-intersecting? Two types of systems spring to mind, though, analogously to the hypergraphs $\mathcal{F}_0, \mathcal{F}_1, \ldots, \mathcal{F}_{r-l}$ described in §7, we can think of many more. First, take the system $\mathcal{F} = \{A \subset X : L \subset A\}$, where L is a fixed l-subset of X. Clearly $\mathcal{F}$ is l-intersecting and $|\mathcal{F}| = 2^{n-l}$. This construction corresponds to $\mathcal{F}_0$ in §7 and to $X^{(r)}(L_1)$ in Theorem 12.3.

To construct another large l-intersecting family, suppose first that $n + l$ is even, say $n + l = 2k$. For $\mathcal{F}$ take $X^{(\geq k)} = \bigcup_{i=k}^{n} X^{(i)}$, the set of all subsets of size at least k. Then for $A, B \in \mathcal{F}$ we have $|A \cap B| = |A| + |B| - |A \cup B| \geq k + k - n = l$ so $\mathcal{F}$ is indeed l-intersecting. Clearly $|\mathcal{F}| = \sum_{i=k}^{n} \binom{n}{i}$. If $n + l$ is odd, say $n + l = 2k - 1$, then a slight enlargement of $X^{(\geq k)}$ will be an l-intersecting family, namely $\mathcal{F} = X^{(\geq k)} \cup (X \setminus \{x\})^{(k-1)}$, where $x \in X$ is an arbitrary element. Note that $(X \setminus \{x\})^{(k-1)}$ is an l-intersecting $(k-1)$-graph so $\mathcal{F}$ is l-intersecting and $|\mathcal{F}| = \sum_{i=k}^{n} \binom{n}{i} + \binom{n-1}{k-1}$.

It turns out that the second construction is better than the first and is, in fact, best possible. This was first proved by Katona (1964) and our main aim in this section is to present Kleitman's (1966b) proof of this result.

Theorem 4. *Let $\mathcal{F} \subset \mathcal{P}(X)$ be an l-intersecting family. Then if $n + l = 2k$ we have*

$$|\mathcal{F}| \leq |X^{(\geq k)}| = \sum_{i=k}^{n} \binom{n}{i}$$

and if $n + l = 2k - 1$ then

$$|\mathcal{F}| \leq |X^{(\geq k)} \cup (X - \{x\})^{(k-1)}| = \sum_{i=k}^{n} \binom{n}{i} + \binom{n-1}{k-1},$$

where $x \in X$.

Both inequalities are best possible. ∎

We shall prove a reformulation of Theorem 4 which deserves attention in its own right. Note that l-intersecting families are closely tied to families in which no two sets have a large union. Indeed, for $A, B \subset X$ we have

$$|(X \setminus A) \cup (X \setminus B)| = |X| - |A \cap B|$$

so $\mathcal{F}$ is l-intersecting iff $\mathcal{F} = \{X \setminus A : A \in \mathcal{F}\}$ is such that $|A \cup B| \leq n - l$ for all $A, B \in \mathcal{F}^c$. If $\mathcal{F} = X^{(\geq k)}$ then $\mathcal{F}^c = X^{(\geq n-k)}$ and if $\mathcal{F} = X^{(\geq k)} \cup (X \setminus \{x\})^{(k-1)}$ then $\mathcal{F}^c = X^{(\leq n-k)} \cup X_x^{(n-k+1)}$. Therefore Theorem 4 is equivalent to the following result.

Theorem 5. *Let $\mathcal{F} \subset \mathcal{P}(X)$ be such that $|A \cup B| \leq m < n$ for all $A, B \in \mathcal{F}$. Then if $m = 2k$ we have*

$$|\mathcal{F}| \leq |X^{(\leq k)}| = \sum_{i=0}^{k} \binom{n}{i} \tag{3}$$

and if $m = 2k + 1$ then

$$|\mathcal{F}| \leq |X^{(\leq k)} \cup X_x^{(k+1)}| = \sum_{i=0}^{k} \binom{n}{i} + \binom{n-1}{k}. \tag{4}$$

Both inequalities are best possible.

Proof. The set systems $X^{(\leq k)}$ and $X^{(\leq k)} \cup X_x^{(k+1)}$ show that the inequalities, if true, are best possible.

The main step in the proof of the inequalities (3) and (4) is that, as in the proof of the Kruskal-Katona theorem, we may assume that $\mathcal{F}$ is left compressed. Before stating a lemma implying this, let us recall some relevant definitions. For $1 \leq i \leq n$, $1 \leq j \leq n$, $i \neq j$, the operator $R_{ij} : \mathcal{P}(X) \to \mathcal{P}(x)$ replaces the element j by the element i, whenever possible:

$$R_{ij}(A) = \begin{cases} (A \setminus \{j\}) \cup \{i\} & \text{if } j \in A \text{ and } i \notin A \\ \text{A} & \text{otherwise} \end{cases}$$

Furthermore, for $\mathcal{F} \subset \mathcal{P}(X)$ we have

$$\tilde{R}_{ij}(\mathcal{F}) = \{R_{ij}(A) : A \in \mathcal{F}\} \cup \{A : A \in \mathcal{F} \text{ and } R_{ij}(A) \in \mathcal{F}\}.$$

We know (and it is immediate) that $\tilde{R}_{ij}$ is a one-to-one map, mapping $X^{(r)}$ onto $X^{(r)}$ and $|\tilde{R}_{ij}(\mathcal{F})| = |\mathcal{F}|$ for all $|\mathcal{F} \subset \mathcal{P}(X)$.

Lemma 6. *Let $|A \cup B| \leq m$ for all $A, B \in \mathcal{F}$. Then*

$$|A \cup B| \leq m \text{ for all } A, B \in \tilde{R}_{ij}(\mathcal{F}). \tag{5}$$

Proof. If $R_{ij}(A) \neq A$ and $R_{ij}(B) \neq B$ then

$$|R_{ij}(A) \cup R_{ij}(B)| = |R_{ij}(A \cup B)| = |A \cup B|$$

if $A, R_{ij}(A) \in \mathcal{F}$, $B \in \mathcal{F}$ then

$$|R_{ij}(A) \cup R_{ij}(B)| \leq |A \cup B| \leq m.$$

Hence (5) holds. ∎

A set system $\mathcal{F} \subset \mathcal{P}(X)$ is *left compressed* if $\tilde{R}_{ij}(\mathcal{F}) = \mathcal{F}$ for all $1 \leq i \leq j \leq n$. Thus $\mathcal{F}$ is left compressed if $R_{ij}(A) \in \mathcal{F}$ whenever $A \in \mathcal{F}$

and $i < j$. Equivalently, $\mathcal{F}$ is left compressed if whenever $A \in \mathcal{F}$, $A_0 = \{a_1, a_2, \ldots, a_s\} \subset A$, $a_1 > a_2 > \ldots > a_s$, $B_0 = \{b_1, b_2, \ldots, b_s\} \subset X \setminus A$, $b_1 > b_2 > \ldots b_s$ and $a_1 > b_1$, $a_2 > b_2$, ..., $a_s > b_s$, the set $(A \setminus A_0) \cup B_0$ belongs to $\mathcal{F}$. Indeed, $(A \setminus A_0) \cup B_0$ is precisely $R_{b_1 a_1} R_{b_2 a_2} \ldots R_{b_s a_s}(A)$. (Note that the elements of A_0 and B_0 are enumerated in decreasing order!)

By imitating the proof of Lemma 5.2, we find that Lemma 6 implies the following assertion.

Lemma 7. *If $|A \cup B| \leq m$ for all $A, B \in \mathcal{F}$ then there is a left compressed set system $\mathcal{F}' \subset \mathcal{P}(X)$ such that $|\mathcal{F}'| = |\mathcal{F}|$ and $|A \cup B| \leq m$ for all $A, B \in \mathcal{F}'$.* ■

Let us turn to the proof of inequalities (3) and (4). Let $\mathcal{F} \subset \mathcal{P}(X)$ be such that $|A \cup B| \leq m$ for all $A, B \in \mathcal{F}$. We may assume that $\mathcal{F}$ is a maximal system satisfying this condition and so $\mathcal{F}$ is monotone decreasing: $A \in \mathcal{F}$ and $B \subset A$ imply $B \in \mathcal{F}$. More importantly, by Lemma 7 we may assume that $\mathcal{F}$ is left compressed.

First let us consider the case $m = 2k$. As usual, write $\mathcal{F}_i = X^{(i)} \cap \mathcal{F}$, $i = 0, 1, \ldots, n$. Then $\mathcal{F} = \sum_{i=0}^{m} \mathcal{F}_i$ since $\mathcal{F}_i = \emptyset$ for $i > m$. For each r, $1 \leq r \leq k$, we shall define a 1–1 map $\mathcal{F}_{k+r} \to X^{(k+1-r)} \setminus \mathcal{F}_{(k+1-r)}$. Having done this, inequality (3) follows since then

$$|\mathcal{F}_{k+r}| + |\mathcal{F}_{k+1-r}| \leq \binom{n}{k+1-r}$$

and so

$$|\mathcal{F}| = |\mathcal{F}_0| + \sum_{r=1}^{k} |\mathcal{F}_{k+r}| + |\mathcal{F}_{k+1-r}| \leq \sum_{i=0}^{k} \binom{n}{i}.$$

Consider a set $A \in \mathcal{F}_{k+r}$, $r \geq 1$. What $(k+1-r)$-element set should we associate with A? We shall choose one that is disjoint from A and so is not in $\mathcal{F}_{k+1-r}$, and is closely related to A. Suppose $A = \{a_1, a_2, \ldots, a_{k+r}\}$, where $a_1 > a_2 > \ldots > a_{k+r}$. Form a sequence $b_1, b_2, \ldots, b_s$ as follows. Set $B_1 = [a_1] \setminus A = \{1, 2, \ldots, a_1\} \setminus A$. If $B_1 \neq \emptyset$, take the empty sequence $(s = 0)$, otherwise set $b_1 = \max B_1$ and continue the process. Having defined $b_1 > b_2 > \ldots > b_t$, set $B_{t+1} = [a_{t+1}] \setminus (A \cup \{B_1, b_2, \ldots, b_t\})$. If $B_{t+1} = \emptyset$, terminate the sequence $(s = t)$, otherwise let $b_{t+1} = \max B_{t+1}$. Thus $b_1, b_2, \ldots, b_s$ are the s largest elements not in A into which $a_1, a_2, \ldots, a_s$ can be mapped: $a_1 \to b_1, a_2 \to b_2, \ldots, a_s \to b_s$ such that $a_i > b_i$ for each i.

Set $B_0(A) = \{b_1, b_2, \ldots, b_s\}$ and let $B_1(A)$ be the set of $k+1-r-s$ largest elements of $X \setminus (A \cup B_0(A))$. Note that $k+1-r-s \geq 1$ since with $A_0 = \{a_1, a_2, \ldots, a_s\}$ we have $A_0 \subset A$, $B_0 \subset X \setminus A$, $a_1 > b_1, a_2 > b_2, \ldots, a_s > b_s$ and so, as $\mathcal{F}$ is left compressed, $(A \setminus A_0) \cup B_0 \in \mathcal{F}$ implying $|(A \setminus A_0) \cup B_0 \cup A| \geq |A \cup B_0| = k+r+s \leq 2k$. Let $B(A) = B_0(A) \cup B_1(A)$. Then $B(A)$ is a $(k+1-r)$-element subset of X, disjoint from A, so $B(A) \notin \mathcal{F}$. We shall send A into $B(A)$.

As an example, take $n = 20$, $k = 6$, $r = 2$ and $A = \{1,2,3,5,6,7,18,20\}$. Then $A_1 = 20 \to b_1 = 19$, $a_2 = 18 \to b_2 = 17$, $a_3 = 7 \to b_3 = 4$ and here the sequence stops, so $s = 3$, $B_0(A) = \{4,17,19\}$, $B_1(A) = \{15,16\}$ and $B(A) = \{4,15,16,17,19\}$.

Is the map $A \to B(A)$ an injection? The parameter s is precisely the maximal number of elements of $B(A)$ which can be put into one-to-one correspondence with elements in $X \setminus B(A)$ and larger than themselves. Hence $B(A)$ determines s. Furthermore, $B_0(A)$ is the largest s-subset of $B(A)$ in the colex order that can be shifted to the right and out of $B(A)$. Thus, in the example above, $19 \to 20$, $17 \to 18$, 16 and 15 cannot be shifted, $4 \to 14$, so $B_0(A) = \{4,17,19\}$. Finally, $B_0(A)$ determines A since A has s elements as little to the right of the elements of $B_0(A)$ as possible. In our example, A must contain 20, then 18, then 5, and $6+2-3=5$ more elements: $1,2,3,6$ and 7.

The argument above shows that $A \to B(A)$ is a one-to-one map from $\mathcal{F} \setminus X^{(\leq k)}$ into $X^{(\leq k)}\mathcal{F}$. Hence $|\mathcal{F}| \leq |X^{(\leq k)}| = \sum_{i=0}^{k} \binom{n}{i}$, proving (3).

Let us give another illustration of the argument that $B(A)$ determines A. Once again, let $n = 20$ and $k = 6$. Suppose $B(A) = \{5,8,17,18\}$. What is A? First of all, $|B(A)| = 6 - r + 1 = 4$ so $r = 3$ and we are looking for a set A with $6+3=9$ elements. Next, $18 \to 20$, $17 \to 19$, $8 \to 16$ and $5 \to 15$ is a permissible right shift so $s = 4$ and $B_0(A) = B(A) = \{5,8,17,18\}$. Finally, $18 \to 19$, $17 \to 20$, $8 \to 9$ and $5 \to 6$ give us four elements to which we must add five more: $1,2,3,4$, and 7. Thus $A = \{1,2,3,4,6,7,9,19,20\}$. As a check, note that, starting with A, $20 \to 18$, $19 \to 17$, $9 \to 8$, $7 \to 5$ is all we can do so $B_0(A) = \{5,8,17,18\}$ and $s = 4$. We need $k-r+1-s$ more elements for $B(A)$. As $k-r+1-s = 6-3+1-4 = 0$, we have $B(A) = B_0(A)$.

The case $m = 2k+1$ needs a little more care. The main argument is as earlier: for $r \geq 1$ define a one-to-one map $\mathcal{F}_{k+1+r} \to X^{(k+1-r)} \setminus \mathcal{F}_{k+1-r}$ by sending A to $B(A)$. Setting $j = k+1+r$, the system $\mathcal{F}_j$ is mapped into $X^{(m+1-j)} \setminus \mathcal{F}_{m+1-j}$, precisely as before. This shows that

$$|\mathcal{F}| \le |\mathcal{F}_{k+1}| + \sum_{i=0}^{k} \binom{n}{i}. \tag{6}$$

To complete the proof, we need a better bound on $\mathcal{F}_{k+1}$ than the trivial estimate $|\mathcal{F}| \le \binom{n}{k+1}$. Fortunately, this is easily obtained, since $\mathcal{F}_{k+1}$ is an intersecting $(k+1)$-graph: if $A, B \in \mathcal{F}_{k+1}$ then $A \cap B \ne \emptyset$ since $|A \cup B| \le m = |A| + |B| - 1$. Hence, by Theorem 7.1 (or by Theorem 2), $\mathcal{F}_{k+1}| \le \binom{n-1}{k}$ and so, by (6), inequality (4) holds as required. ∎

Theorem 2 implies a beautiful result about the metric space $\mathcal{P}(X)$ of all subsets, with the metric $d(A, B) = |A \triangle B|$. This result, due to Kleitman (1966b), is a slight extension of Theorem 5. The *diameter* of a set system $\mathcal{F} \subset \mathcal{P}(X)$ is defined to be diam $\mathcal{F} = \max\{d(A, B) : A, B \in \mathcal{F}\}$.

Theorem 8. *Let $\mathcal{F} \subset \mathcal{P}(X)$ be such that diam $\mathcal{F} \le m < n$. Then if $m = 2k$ we have*

$$|\mathcal{F}| \le |X^{(\le k)}| = \sum_{i=0}^{k} \binom{n}{i} \tag{7}$$

and if $m = 2k + 1$ then

$$|\mathcal{F}| \le |X^{(\le k)} \cup X_x^{(k=1)}| = \sum_{i=0}^{k} \binom{n}{i} + \binom{n-1}{k}. \tag{8}$$

Both inequalities are best possible.

Proof. Since $|A \triangle B| \le |A \cup B|$, if $|A \cup B| \le m$ for all $A, B \in \mathcal{F}$ then diam $\mathcal{F} \le m$. Hence inequalities (7) and (8) are stronger than (3) and (4); in particular, if true, they are best possible.

To deduce Theorem 8 from Theorem 5, for $1 \le h \le n$ and $A \subset X$ define $T_h(A) = A \setminus \{h\}$. Furthermore, let $\tilde{T}_h$ be the analogue of $\tilde{R}_{ij}$, sending a set system $\mathcal{F}$ into a set system $\tilde{T}_h(\mathcal{F})$ of the same cardinality:

$$\tilde{T}_h(\mathcal{F}) = \{T_h(A) : A \in \mathcal{F}\} \cup \{A : A \in \mathcal{F}, T_h(A) \in \mathcal{F}\}.$$

It is immediate that the analogue of Lemma 4 holds:

$$\text{diam } \tilde{T}_h(\mathcal{F}) \le \text{diam } \mathcal{F} \tag{9}$$

for all $\mathcal{F} \subset \mathcal{P}(X)$.

Let now $\mathcal{F} \subset \mathcal{P}(X)$ be a set system of diameter at most m. Then $\mathcal{F}_0 = \tilde{T}_1(\tilde{T}_2(\ldots \tilde{T}_n(\mathcal{F})\ldots))$ satisfies $|\mathcal{F}_0| = |\mathcal{F}|$ and, by (9), diam $\mathcal{F}_0 \leq$ diam $\mathcal{F} \leq m$. Since $\tilde{T}_h(\mathcal{F}_0) = \mathcal{F}_0$ for all h, $1 \leq h \leq n$, $\mathcal{F}_0$ is an ideal: $A \in \mathcal{F}_0$ and $A' \subset A$ imply $A' \in \mathcal{F}_0$.

How large can $|A \cup B|$ be for $A, B \in \mathcal{F}_0$? Set $A' = A \backslash B$. Then $A' \in \mathcal{F}_0$ and $A \cup B = A' \triangle B$. Hence $|A \cup B| \leq m$. Therefore, by Theorem 5, $|\mathcal{F}| = |\mathcal{F}_0|$ satisfies the appropriate inequality. ∎

Let us state an outstanding result of Frankl and Rödl (1986) vaguely related to the theorems above. Suppose $\mathcal{F} \subset \mathcal{P}(X)$ is such that if $A, B \in \mathcal{F}$ and $A \neq B$, then $|A \cap B| \neq \lfloor n/4 \rfloor$. Then $|F| < (2-\epsilon)^n$ where $\epsilon > 0$ is an absolute constant.

Theorem 8 determines the maximal cardinality of a set system of diameter m. In the case $m = 2k$ the result has a rather pleasing reformulation. For $A \in \mathcal{P}(X)$ and $0 \leq k \leq n$ denote by $\mathcal{B}(A, k)$ the *ball of radius k and centre A*:

$$\mathcal{B}(A, k) = \{B \in \mathcal{P}(X) : d(A, B) = |A \triangle B| \leq k\}.$$

Thus $\mathcal{B}(\emptyset, k) = X^{(\leq k)}$. For $k \leq n/2$ the diameter of a ball of radius k is precisely $2k$ (for $k > n/2$ the diameter is n as diam $\mathcal{P}(X) = \max\{|A \triangle B| : A, B \in \mathcal{P}(X)\} = n$). The metric space $(\mathcal{P}(X), d)$ is homogeneous in the sense that if $A, B, A', B' \in \mathcal{P}(X)$ and $d(A, B) = d(A', B')$ then there is a distance-preserving map $\phi : \mathcal{P}(X) \to \mathcal{P}(X)$ such that $\phi(A) = A'$ and $\phi(B) = B'$. In particular, any two balls of radius k are isometric and so have the same number of points (sets).

Theorem 8 claims that a system of diameter $2k < n$ contains at most as many sets as a ball of radius k. In §16 we shall see that balls are also solutions of the isoperimetric problem for sets: the boundary of a set system with $|\mathcal{B}(A, k)|$ sets is at least as large as the boundary of the ball $\mathcal{B}(A, k)$.

In §19 we shall return briefly to intersecting families. Using methods rather different from those we have employed so far, we shall prove Kleitman's theorem stating that a union of k intersecting families contains at most $2^n - 2^{n-k}$ sets. This result is the analogue of Theorem 12.3 and is a substantial extension of the trivial bound 2^{n-1} on the size of an intersecting family.

Exercises

1. Deduce from Theorem 2 the following theorem of Brace and Daykin (1972). Suppose $\mathcal{A}$ is a Sperner system on X such that if $A, B \in \mathcal{A}$

then $A \cap B \neq \emptyset$ and $A \cup B \neq X$. Then

$$|\mathcal{A}| \leq \binom{n-1}{\lfloor (n-2)/2 \rfloor}.$$

(Let $\mathcal{A}^c = \{X \setminus A : A \in \mathcal{A}\}$ and consider the system $\mathcal{A} \cup \mathcal{A}^c$.)

2. (Harder.) Choosing an appropriate weight function for the intervals of T, prove the following theorem of Greene, Katona and Kleitman (1976). If $\mathcal{A} = \mathcal{A}_1 \cup \mathcal{A}_2$ is an intersecting Sperner system on X such that $\mathcal{A}_1 \subset X^{(\leq n/2)}$ and $\mathcal{A}_2 \subset X^{(>n/2)}$ then

$$\sum_{A \in \mathcal{A}_1} \binom{n}{|A|-1}^{-1} + \sum_{A \in \mathcal{A}_2} \binom{n}{|A|}^{-1} \leq 1.$$

3. Let $\mathcal{F} \subset X^{(\leq r)}$ be an intersecting family such that $\bigcap_{F \in \mathcal{F}} F = \emptyset$. Prove that there is a set Y such that $|Y| \leq 3r - 3$ and $|F \cap Y| \geq 2$ for all $F \in \mathcal{F}$. (Let $F_1 \in \mathcal{F}$. If $|F_1 \cap F| \geq 2$ for all $F_1 \in \mathcal{F}$ then $Y = F_1$ will do. Otherwise there is an $F_2 \in \mathcal{F}$ such that $|F_1 \cap F_2| = \{x\}$ for some x. Pick a set $F_3 \in \mathcal{F}$ not containing x and take $Y = F_1 \cup F_2 \cup F_3$.)

4. For $s, t \geq 1$ let $f(s,t)$ be the minimal integer m such that if $\mathcal{A} \subset X^{(\leq s)}$ and $\mathcal{B} \subset X^{(\leq t)}$ are such that $A \cap B \neq \emptyset$ for all $A \in \mathcal{A}$ and $B \in \mathcal{B}$, then there is a set M of size m satisfying $A \cap B \cap M \neq \emptyset$ for all $A \in \mathcal{A}$ and $B \in \mathcal{B}$. Imitate the proof in Exercise 12.3 to show that

$$f(s,t) \leq (s+t-1)\binom{s+t-2}{s-1}.$$

5. Prove that $f(s,t) \leq s + sf(s, t-1)$.
(Follow Ehrenfeucht and Mycielski (1974): fix $A_0 \in \mathcal{A}$ and for $x \in A_0$ set $\mathcal{A}_x = \{A \in \mathcal{A} : x \notin A\}$ and $\mathcal{B}(X) = \{B \setminus \{x\} : B \in \mathcal{B}, x \in B\}$. Find a set M_x for A_x and B_x and set $M = \{A_0 \cup \bigcup \{M_x : x \in A_0\}\}$.)

6. Prove the following theorem of Ehrenfeucht and Mycielski (1974). Let $\mathcal{A}_1, \ldots, \mathcal{A}_k$ be infinite families of sets of size at most s such that if $A_i \in \mathcal{A}_i$ for $i = 1, \ldots, k$, then $A_i \cap A_j \neq \emptyset$ for some $i \neq j$. Then there is a finite set M such that whenever $A_i \in \mathcal{A}_i$ for $i = 1, \ldots, k$ then $A_i \cap A_j \neq \emptyset$ for some $i \neq j$.
(Prove it first for finite families by considering the families $\mathcal{A} = \{\bigcup_1^{k-1} A_i : A_i \in \mathcal{A}_i, A_i \cap A_j = \emptyset \text{ if } i \neq j\}$ and $\mathcal{B} = \mathcal{A}_k$. Extend the result to infinite families as in Exercise 12.5.)

§14. FACTORIZING COMPLETE HYPERGRAPHS

On p. 48 of the Lady's and Gentleman's Diary, 1850, the Rev. Thomas P. Kirkman posed the following puzzle. "Fifteen young ladies in a school walk out three abreast for seven days in succession: it is required to arrange them daily, such that no two shall walk twice abreast." Such an arrangement looks feasible because the fifteen girls can be divided into groups of three and, as a girl has two different companions every day, after seven days she may have been in the same row as each of the other $7 \times 2 = 14$ girls, exactly once with each. It is rather easy to ensure that some of the girls mix appropriately with all the others but we do run into difficulties in finding an arrangement satisfying all the conditions. The enterprising reader is encouraged to stop here and give the problem a try.

As Kirkman wrote (1850c), the puzzle excited "some attention among a far higher class of readers than those for whom the first 48 pages of the Diary are intended". (This is not an exaggeration: Kirkman's puzzle was one of six queries, two others being "What is the cause of the contraction of hemp and catgut strings in a damp atmosphere?" and "Required the origin of the custom of making fools on the first day of April".) Several solutions of the puzzle were published, among others, by Cayley (1850) and Kirkman (1850b). In his solution, Cayley (1850) mentioned an extension of the puzzle, due to Sylvester, "to make the school walk out every day in the quarter so that each three may walk together". To be precise, suppose that the 15 schoolgirls go for walks during each day of 13 weeks. Is it possible for any three girls to be in the same row in precisely one of 91 walks? Once again, the choice of 91 is judicious: there are $\binom{15}{3} = 5 \times 91$ triples and each walk takes care of 5 of those. Putting it another way, each girl must be in the same row precisely once with each of her $\binom{14}{2} = 91$ pairs of classmates. Kirkman (1850c) proved that such a schedule was indeed possible.

Let us formulate the extension of this second problem to complete r-graphs. A *1-factor* of a hypergraph $H = (X, \mathcal{E})$ is a set of edges $\mathcal{E}' \subset \mathcal{E}$ such that every vertex of X belongs to precisely one edge in $\mathcal{E}'$. A *1-factorization* or simply *factorization* of H is a partition of the edge-set into 1-factors: $\mathcal{E} = \mathcal{F}_1 \cup \mathcal{F}_2 \cup \ldots \cup \mathcal{F}_l$, where every $\mathcal{F}_i$ is a 1-factor and $\mathcal{F}_i \cap \mathcal{F}_j = \emptyset$ for $1 \leq i < j \leq l$. Thus Sylvester's Problem asks for a factorization of $K_{15}^{(3)}$. In this section we shall discuss 1-factorizations of $K_n^{(r)} = (X, X^{(r)})$, the complete r-graph of order n.

If an r-graph $H = (X, \mathcal{E})$ has a 1-factor then $r|n$ so if $K_n^{(r)}$ is factorizable then $r|n$. (As is customary, for integers a and b, $a|b$ means that a divides the integer b). The reader probably expects this trivial necessary condition to be sufficient as well and, had he not been alerted by Kirkman's Fifteen Schoolgirls Problem, he might even expect the proof to be rather straightforward. As it happens, the trivial necessary condition $r|n$ is indeed sufficient for the existence of a factorization but the proof is far from simple. In fact, though in a somewhat vague form, the assertion has been around for over a hundred years, starting with Sylvester's Problem, a complete solution was given only recently by Baranyai (1975). The interested reader should attempt to prove the result himself so as to get a feel for the difficulty.

Theorem 1. *If $r|n$ then $K_n^{(r)}$ is 1-factorizable.* ∎

As in this book we are restricting ourselves to fairly manageable calculations, we shall not give an entire proof; we shall be satisfied with presenting some preliminary results which give the flavour of the proof. These results also lead to an extension of Theorem 1, stating that a complete k-partite r-graph $K_{k\times m}^{(r)}$ is factorizable if $r|km$. Here $K_{k\times m}^{(r)}$ is the r-graph with vertex set $X = \bigcup_{i=1}^{k} X_i$, $|X| = n = km$, $|X_i| = m$, $1 \leq i \leq k$, and edge set $E(K_{k\times m}^{(r)}) = \{E \in X^{(r)} : |E \cap X_i| \leq 1, 1 \leq i \leq k\}$.

Before getting down to some work, let us point out that for $r = 2$, i.e. for the more usual graphs (edge graphs), Theorem 1 is very simple indeed. To get a factorization of $K^{2k} = K_{2k}^{(2)}$, the complete edge graph of order $2k \geq 4$, consider the vertex set $W = \{x_1, \ldots, x_{2k-1}\}$ of a regular $(2k-1)$-gon and take another vertex x_{2k}. To get a 1-factor of the complete graph on $V = \{x_1, \ldots, x_{2k}\} = W \cup \{x_{2k}\}$, join all pairs of points in W which determine parallel lines. Precisely one point in W will not be joined to any point in W; join it to x_{2k} (see Figure 1). Repeating the construction for every direction determined by two points of W, we get an appropriate factorization. This construction is very similar to the customary factorization of a complete graph into Hamilton cycles; see Bollobás (1979, p. 13).

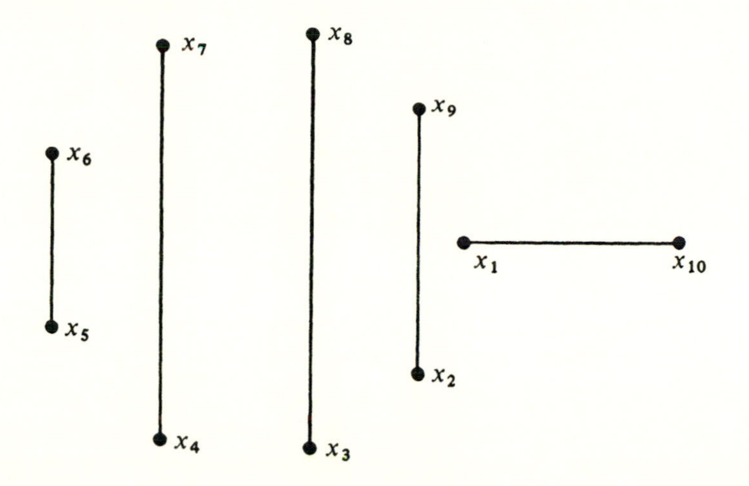

Figure 1. *A 1-factor of K^{10}; the rotations give a factorization*

More formally, to get a factorization $F_1 \cup \ldots F_{2k-1}$ of K^{2k}, with vertex set $[2k] = \{1, 2, \ldots, 2k\}$, let F_l consist of all k–1 edges ij satisfying $1 \le i < j \le 2k-1$ and $i + j \equiv l \pmod{2k-1}$, together with the edge joining h to $2k$ for $2h \equiv l \pmod{2k-1}$.

Let us turn to the results used in the proof of Theorem 1 and its extension. Let $I_1, I_2, \ldots, I_k$ be finite sets and set $I = I_1 \times \ldots \times I_k$. Call a set $J \subset I$ a *face* of I if J consists of all $\underline{i} = (i_1, i_2, \ldots, i_k) \in I$ for which some (namely $0, 1, \ldots, k-1$ or k) of the elements $i_1, i_2, \ldots, i_k$ are fixed and the others take all possible values. Thus if $I_1 = [3]$, $I_2 = [4]$ and $I_3 = [2]$ then $\{a_{321}, a_{322}\}$, $\{a_{211}, a_{212}, a_{221}, a_{222}, a_{231}, a_{232}, a_{241}, a_{242}\}$ and $\{a_{142}\}$ are all faces of $I = I_1 \times I_2 \times I_3$. If $a_{\underline{i}}$ is a real number for every $\underline{i} \in I$ then $\{a_{\underline{i}} : \underline{i} \in I\}$ is called a *k-dimensional array.* Thus a 2-dimensional array is just a real matrix.

Our first aim is to show that every real matrix can be approximated by an integer matrix. For real numbers x and y the relation $x \approx y$ means that $|x - y| < 1$. Note that if x is a real number and a is an integer then $x \approx a$ iff either $a = \lfloor x \rfloor$, the greatest integer not greater than x, or $x = \lceil x \rceil$, the smallest integer not smaller than x. In particular, if both a and b are integers then $a \approx b$ iff $a = b$.

Theorem 2. *Let $\{\epsilon_{ij} : (i,j) \in I = I_1 \times I_2\}$ be a 2-dimensional* *array. Then there is an integer array* $\{a_{ij} : (i,j) \in I\}$ *such that*

$$\sum_J \epsilon_{ij} \approx \sum_J a_{ij}$$

for every face J *of* I.

Proof. Suppose (ϵ_{ij}) is an n by m matrix. Let us show that we may assume that every row sum is 0 and every column sum is 0. Indeed, set $\beta_i = \sum_{j=1}^m \epsilon_{ij}$, $\gamma_j = \sum_{i=1}^m \epsilon_{ij}$, and attach another row and another column to our matrix:

$$\epsilon_{n+1,j} = -\gamma_j, \epsilon_{i,m+1} = -\beta_i \quad \text{and} \quad \epsilon_{n+1,m+1} = \sum_{i=1}^n \beta_i = \sum_{j=1}^m \gamma_j.$$

In the new matrix every row-sum is 0 and every column-sum is 0. If (a_{ij}) is an $n+1$ by $m+1$ matrix satisfying the conclusions of the theorem then $\{a_{ij} : 1 \le i \le n, 1 \le j \le m\}$ will do for the original (ϵ_{ij}). Indeed, for $1 \le i \le n$,

$$\sum_{j=1}^m a_{ij} = -a_{i,m+1} \approx -(-\beta_i) = \beta_i = \sum_{j=1}^m \epsilon_{ij}$$

and, similarly, for $1 \le j \le m$,

$$\sum_{i=1}^n a_{ij} = -a_{n+1,j} \approx -(-\gamma_j) = \gamma_j = \sum_{i=1}^n \epsilon_{ij}.$$

Why is the sum of all entries well approximated by $\sum_{i=1}^n \sum_{j=1}^m a_{ij}$? Because

$$\sum_{i=1}^n \sum_{j=1}^m \epsilon_{ij} = \sum_{i=1}^n \beta_i = \epsilon_{n+1,m+1}$$

$$\approx a_{n+1,m+1} = \sum_{i=1}^n a_{i,m+1} = \sum_{i=1}^n \sum_{j=1}^m a_{ij}.$$

Suppose then that in the matrix (ϵ_{ij}) every row-sum and every column-sum is 0. Choose a matrix $A = (a_{ij})$ with all row-sums and all column-sums 0, such that

$$\lfloor \epsilon_{ij} \rfloor \le a_{ij} \le \lceil \epsilon_{ij} \rceil \tag{1}$$

for all entries, and as many of the entries a_{ij} are integers as possible. Suppose that, contrary to the assertion of the theorem, not every a_{ij} is an integer.

Call a sequence of entries $a_0, a_1, \ldots, a_{2s-1}, a_{2s} = a_0$ a *circuit* if no a_i is an integer, a_0 and a_1 are in the same row, a_1 and a_2 are in the same column, a_2 and a_3 are in the same row etc. Note that no row of A contains precisely one non-zero entry and no column of A contains precisely one non-zero entry. Therefore, as not every entry of A is an integer, A contains a circuit. Indeed, starting with a non-zero entry a_{ij}, we can find another non-zero entry in the same row, say a_{il}; then we can find another non-zero entry in the l-th column, say a_{hl}; etc. Eventually, we must return to a row or a column previously visited and so we can find a circuit.

For example if, starting with a_{23}, we get the sequence $a_{23}, a_{27}, a_{97}, a_{98}, a_{68}, a_{61}, a_{31}, a_{37}$ then a_{37} is in column 7, which we have visited earlier, and we get the circuit $a_{37}, a_{97}, a_{98}, a_{68}, a_{61}, a_{31}$.

Given a circuit $a_0 a_1 \ldots a_{2s}$, let $\epsilon = \min\{\lceil a_t \rceil - a_t, a_t - \lfloor a_t \rfloor\}$. Replace a_0 by $a_0 + \epsilon$, a_1 by $a_1 - \epsilon$, a_2 by $a_2 + \epsilon$, etc. The new matrix also satisfies (1), has all row-sums and column-sums 0 and has more integer entries than A, contradicting the choice of A. Hence there is an integer matrix $A = (a_{ij})$ satisfying (1), so our result is proved. ∎

In order to factorize hypergraphs one would need Theorem 1 for higher-dimensional arrays. Unfortunately, Theorem 1 cannot even be extended to three-dimensional arrays. The key to the proof of Theorem 1 was the existence of circuits of fractional entries; the analogue of such a circuit need not exist in a three-dimensional array. It is a little surprising that Theorem 1 cannot even be extended to 2 by 2 by 2 arrays. For example, let $E = \{\epsilon_{ijl} : 1 \le i, j, l \le 2\}$ be given by

$$\epsilon_{ijl} = \begin{cases} 1/2 & \text{if } i + j + l \text{ is odd} \\ 0 & \text{if } i + j + l \text{ is even} \end{cases}$$

Suppose $A = (a_{ijl})$ is an integer array approximating E. As $\sum \epsilon_{ijl} = 2$, two entries of A are 1's, the other are 0's. Furthermore, one of the 1's is in the bottom face and the other in the top face, so we may assume that they are $a_{111} = 1$ and $a_{212} = 1$ (see Figure 2). But then $\sum_{i,l} \epsilon_{i2l} = 1 \not\approx \sum_{i,l} a_{i2l} = 0$, contradicting our assumption on A.

Though Theorem 1 cannot be generalized to arbitrary three-dimensional arrays, it can be extended to three-dimensional arrays in a rather restricted form.

$$E_1 = \begin{pmatrix} 1/2 & 0 \\ 0 & 1/2 \end{pmatrix} \qquad E_2 = \begin{pmatrix} 0 & 1/2 \\ 1/2 & 0 \end{pmatrix}$$

$$A_1 = \begin{pmatrix} 1 & 0 \\ 0 & 0 \end{pmatrix} \qquad A_2 = \begin{pmatrix} 0 & 0 \\ 1 & 0 \end{pmatrix}$$

Figure 2. *The matrices $A_1 = (a_{ij1})$, $A_2 = (a_{ij2})$, $E_1 = (\epsilon_{ij1})$ and $E_2 = (\epsilon_{ij2})$*

Theorem 3. *Let $I = I_1 \times I_2 \times I_3$, where I_1, I_2 and I_3 are finite sets and let $\{\epsilon_{ijl} : (i,j,l) \in I\}$ be an array of real numbers such that ϵ_{ijl} is independent of l : $\epsilon_{ijl} = \epsilon_{ijk}$ for all $k, l \in I_3$. Then there is an array of integers $\{a_{\underline{i}} : \underline{i} \in I\}$ such that*

$$\sum_{\underline{i} \in J} a_{\underline{i}} \approx \sum_{\underline{i} \in J} \epsilon_{\underline{i}}$$

for every face J of I.

Proof. Let us apply induction on $|I_3| = m$. For $m = 1$ the assertion is just Theorem 2. Assume then that $m > 1$ and the result holds for smaller values of m. For simplicity let us take $I_3 = [m]$ so that in ϵ_{ijl} the suffix l takes the values $1, 2, \ldots, m$.

Define $\epsilon_{ij} = \sum_{l=1}^{m} \epsilon_{ijl} = m\epsilon_{ijl}$ and apply Theorem 2 to the array $\{\epsilon_{ij} : (i,j) \in I_1 \times I_2\}$, obtaining (a_{ij}). Then set $\epsilon'_{ij} = a_{ij}/m$ and apply Theorem 2 to the array $\{\epsilon'_{ij} : (i,j) \in I_1 \times I_2\}$ to obtain an integer array $\{a_{ijm} : (i,j) \in I_1 \times I_2\}$. (The notation suggests that (a_{ijm}) is the upper layer of the three-dimensional array we are looking for — this is precisely the case.)

Next set $\epsilon_{ijl} = (a_{ij} - a_{ijm})/(m-1)$, $l \in [m-1] = I'_3$, $I' = I_1 \times I_2 \times I'_3$ and apply the induction hypothesis to the array $\{e'_{ijl} : (i,j,l) \in I'\}$ to produce an integer array $\{a_{ijl} : (i,j,l) \in I'\}$. We claim that $\{a_{ijl} : (i,j,l) \in I\}$ has the required properties.

This claim is easily shown with the aid of the following two relations. Let ϵ be a real number, a, b, c, m integers and $m \geq 2$. Then

$$a \approx m\epsilon \quad \text{and} \quad b \approx a/m \quad \text{imply} \quad b \approx \epsilon, \tag{2}$$

$$b \approx a/m \quad \text{and} \quad c \approx (a-b)/(m-1) \quad \text{imply} \quad a/m \approx c. \tag{3}$$

For example, with $\epsilon = \sum_j \epsilon_{ijm}$, $a = \sum_j a_{ij}$ and $b = \sum_j a_{ijm}$ we have $a \approx \sum_j \epsilon_{ij} = m \sum_j \epsilon_{ijm}$ and $b \approx \sum_j \epsilon'_{ij} = \frac{1}{m} \sum_j a_{ij} = a/m$ so, by (2), $\sum_j a_{ijm} \approx \sum_j \epsilon_{ijm}$, as required. The other assertions about $\{a_{ijl} : (i,j,l) \in I\}$ can be checked similarly. ∎

The last result which we present in order to convey the flavour of the proof of Baranyai's main theorem concerns almost constant sequences. The result is used in the proof of the induction step.

Call an integer sequence $(a_i)_{i\in I}$ *almost constant* if $|a_i - a_j| \le 1$ for all $i, j \in I$, i.e., if for some integer b every a_i is b or $b+1$. If the index set I is finite (and in what follows it will always be finite) then $(a_i)_{i\in I}$ is almost constant iff $a_i \approx a$ for every $i \in I$, where $a = \sum_{i\in I} a_i/|I|$ is the average value.

Lemma 4. *Let $(a_i)_{i\in I}$ and $(a_j)_{j\in J}$ be almost constant finite sequences such that $I \cap J = \emptyset$ and*

$$\sum_{i\in I} a_i \approx \frac{|I|}{|I|+|J|} \sum_{i\in I\cup J} a_i. \tag{4}$$

Then $(a_i)_{i\in I\cup J}$ is almost constant.

Proof. Suppose $(a_i)_{i\in I\cup J}$ is not almost constant. Then, by interchanging I and J, if necessary, I and J can be partitioned into non-empty sets: $I = I_1 \cup I_2$, $J = J_1 \cup J_2$, such that

$$\begin{array}{lll} a_i = a & \text{if} & i \in I_1, \\ a_i = a+1 & \text{if} & i \in I_2, \\ a_j = c \ge a+1 & \text{if} & j \in J_1, \\ a_j = c+1 & \text{if} & j \in J_2. \end{array}$$

Setting $\alpha = |I_1|$, $\beta = |I_2|$, $\gamma = |J_1|$ and $\delta = |J_2|$, we find that

$$\frac{|I|}{|I|+|J|} \sum_{i\in I\cup J} a_i = \frac{\alpha+\beta}{\alpha+\beta+\gamma+\delta}\{(\alpha+\beta)a + \beta + (\gamma+\delta)c + \delta\}$$

$$\begin{aligned} &\ge (\alpha+\beta)a + \frac{\alpha+\beta}{\alpha+\beta+\gamma+\delta}(\beta+\gamma+2\delta) \\ &= (\alpha+\beta)a + \beta + \frac{\alpha\gamma+\beta\delta+2\alpha\delta}{\alpha+\beta+\gamma+\delta} \\ &\ge (\alpha+\beta)a + \beta + 1 = \sum_{i\in I} a_i + 1, \end{aligned}$$

contradicting (4). ∎

Theorems 2 and 3 and Lemma 4, together with further ingenious arguments including a rather complicated induction step, enabled Baranyai (1979) to prove not only Theorem 1, but also a considerable extension of it. We shall not reproduce the argument but only some of the more important consequences of this extension. Recall that the *degree* $d(x)$ of a vertex $x \in X$ of a hypergraph is the number of edges containing x. A hypergraph is *regular* if it is d-regular for some d and it is *almost regular* if the degrees of any two vertices differ by at most 1, i.e., if $d(x)$ is an almost constant function on the vertex set.

A *factorization* of a hypergraph $(X, \mathcal{E})$ is a partition of the edge set: $\mathcal{E} = \sum_1^s \mathcal{F}_i$; the hypergraphs $(X, \mathcal{F}_i)$ are the *factors* of this factorization. If every factor is d-regular for some d then we have a *d-factorization.*

Let us write $\mathcal{E}$ for the edge set of $K_{k\times m}^{(r)}$, the complete k-partite r-graph on X with m vertices in each class, and set $N = |\mathcal{E}|$. Note that $n = |X| = km$ and

$$N = \binom{k}{r} m^r.$$

As customary, $a|b$ means that a divides b.

Theorem 5. *Let $e_1, e_2, \ldots, e_s$ be non-negative integers such that $\sum_1^s e_i = N$. Then $\mathcal{E}$ can be partitioned into s sets: $\mathcal{E} = \bigcup_1^s \mathcal{E}_i$ such that $|\mathcal{E}_i| = e_i$ and every $(X, \mathcal{E}_i)$ is an almost regular r-graph.* ∎

Corollary 6. *The complete k-partite r-graph $K_{k\times m}^{(r)}$ is d-factorizable iff $r|dn$ and $dn|rN$.*

Proof. Suppose that $K_{k\times m}^{(r)}$ has a d-factorization. Then each factor must have dn/r edges. Hence dn/r is an integer and this integer divides N.

Conversely, suppose $a = dn/r$ is an integer and so is $s = N/a$. Set $a_1 = a_2 = \ldots = a_s = a$ and apply Theorem 5 to obtain a d-factorization. ∎

Corollary 7. *The graph $K_{k\times m}^{(r)}$ is d-factorizable, where $d = r/(r, n)$ and (r, n) is the greatest common divisor of r and n. In particular, if $r|n = km$ then $K_{k\times m}^{(r)}$ is 1-factorizable.*

Proof. Since $dn = rn/(r,n)$ is the least common multiple of r and n, it is divided by r. Furthermore,

$$rN = r\binom{k}{r}m^r = k\binom{k-1}{r-1}m^r = n\binom{k-1}{r-1}m^{r-1}$$

is divisible by r and n so it is divisible by dn. Hence the conditions of Corollary 6 are satisfied so $K^{(r)}_{k\times m}$ is d-factorizable. ∎

Note that Theorem 1 is a special case of Corollary 7: if $m = 1$ then $K^{(r)}_{k\times m} = K^{(r)}_n$ is 1-factorizable.

To conclude this section, let us say a few words about some related problems. A ***Steiner triple system*** (STS) is a set of triples of a set X such that every pair of elements is contained in precisely one of these triples. Thus an STS is a 3-graph $(X, \mathcal{E})$ such that for all pairs $\{a,b\} \in X^{(2)}$ there is precisely one edge $E \in \mathcal{E}$ containing $\{a,b\}$. Suppose there is an STS of order n, i.e. one on a set X with $|X| = n$. Then $|X^{(2)}| = \binom{n}{2}$ must be divisible by 3 so $6|n(n-1)$. Furthermore, n must be odd, since if $x \in X$ is in t triples then $n = 2t + 1$. Hence either $n = 6k + 1$ or $n = 6k + 3$.

In 1853 Steiner posed a sequence of questions (see Ex. 7), the simplest of which was whether one can have an STS for all n of the form $6k + 1$ and $6k + 3$. Unknown to Steiner, this problem had been solved earlier by Kirkman (1847), who proved that the answer is in the affirmative: if $n \equiv 1$ or $3 \pmod 6$ then there is a Steiner triple system of order n. (Thus the accepted terminology is not too appropriate! One must admit, though, that Kirkman read his paper before the Literary and Philosophical Society of Manchester.) In Exercise 5 we give an elegant construction for Steiner triple systems of prime order $p \equiv 1 \pmod 6$.

A ***Kirkman triple system*** (KTS) *of order n* is a Steiner triple system of order n whose $n(n-1)/6$ triples are partitioned into $(n-1)/2$ classes such that each class consists of $n/3$ disjoint triples, i.e. each class is a 1-factor. Thus Kirkman's Fifteen Schoolgirls Problem asks precisely for a KTS of order 15. If there is a KTS of order n then, as $(n-1)/2$ and $n/3$ are both integers, we must have $n = 6k+3$ for some integer k. For over a hundred years it was unknown whether there exists a KTS for $n \equiv 3 \pmod 6$. It was a major achievement when in 1971 Ray-Chaudhuri and Wilson proved that there is indeed a Kirkman triple system of order $n = 6k + 3$ for every k, $k = 0, 1, \ldots$.

Exercises

1. Let $\{\epsilon_{ijl} : 1 \leq i,j,l \leq 2\}$ be given by

$$\epsilon_{ijl} = \begin{cases} 1/2 & \text{if } i+j+l=4 \\ 0 & \text{otherwise} \end{cases}$$

Show that there is no integer array $A = \{a_{ijl} : 1 \leq i,j,l \leq 2\}$ such that $\sum_J c_{ijl} \approx \sum_J \epsilon_{ijl}$ for every face J.

2. Note that in relation (2) we assumed that a is $\lfloor m\epsilon \rfloor$ or $\lceil m\epsilon \rceil$, b is $\lfloor a/m \rfloor$ or $\lceil a/m \rceil$ and concluded that b is $\lfloor \epsilon \rfloor$ or $\lceil \epsilon \rceil$. Can we pin down which of these two possible values b must take? Show that

$$\lfloor \lfloor m\epsilon \rfloor /m \rfloor = \lfloor \epsilon \rfloor \qquad \text{and} \qquad \lceil \lceil m\epsilon \rceil /m \rceil = \lceil \epsilon \rceil$$

but both $\lceil \lfloor m\epsilon \rfloor /m \rceil$ and $\lfloor \lceil m\epsilon \rceil /m \rfloor$ can be either $\lfloor \epsilon \rfloor$ or $\lceil \epsilon \rceil$.

3. Show that the following extension of (3) is false: if $b \approx a/m$ then $b \approx (a-b)/(m-1)$.

4. Suppose G is an almost regular r-graph on X having m edges. Prove that G is regular iff $n|mr$.

5. (Harder.) Prove that the following construction gives a Steiner triple system of prime order $p \equiv 1 \pmod 6$. Let a be a primitive element of the field $\mathbf{Z}_p$, the field of integers modulo $p = 6s+1$. Working in $\mathbf{Z}_p$, first take the following s *initial* triples: $\{a^i, a^{2s+i}, a^{4s+i}\}, i = 0, 1, \ldots, s-1$, and then, for our STS on $\mathbf{Z}_p$, simply take all translates of these initial triples, i.e. take all triples of the form $\{a^i + b, a^{2s+i} + b, a^{4s+i} + b\}$, $i = 0, 1, \ldots, s-1$ and $b \in \mathbf{Z}_p$. Thus, for $p = 13$ we may take $a = 2$, then we obtain the initial triples $\{1,3,9\}$ and $\{2,6,5\}$, and so, writing 13 instead of 0, we get the following STS: $\{1,3,9\}, \{2,4,10\}, \ldots, \{12,1,7\}$, $\{13,2,8\}; \{2,6,5\}, \{3,7,6\}, \ldots, \{13,4,3\}, \{1,5,4\}$.

(All we have to check is that among the $6s$ differences obtained from the initial triples every non-zero element of our field appears precisely once, i.e. no two elements of the form $\pm(a^i - a^{2s+i})$, $\pm(a^i - a^{4s+i})$, $\pm(a^{2s+i} - a^{4s+i})$ are equal. To show this, starting with $a^{p-1} - 1 = a^{6s} - 1 = 0$, prove that $a^{2s} - a^s + 1 = 0$ in our field.)

6. Let F_k be the 3-graph with vertex set $[k+2]$ and edge set $\{123, 124, \ldots, 12(k+2)\}$. Prove that

$$\lim_{n\to\infty} ex(n; F_k)/n^2 = \frac{k-1}{6}$$

where $ex(n; F)$ is the forbidden graph extremal function defined in §8. (Note that $ex(n; F_2) \le \frac{1}{3}\binom{n}{2}$, with equality iff there is an STS of order n. Similarly, $ex(n; F_k) \le \frac{k-1}{2}\binom{n}{2}$. To find a lower bound, take an n for which there is an STS of order n, say $(X, \mathcal{E}_1)$. Take $k-2$ random permutations of X mapping $\mathcal{E}_1$ into STS's $\mathcal{E}_2, \mathcal{E}_3, \ldots, \mathcal{E}_{k-1}$. Give a lower bound for the expectation of $|\bigcup_1^{k-1} \mathcal{E}_i|$.)

7. Suppose $3 \le k \le n$ and there is a Steiner system $\mathcal{F} = \bigcup_{i=3}^{k} \mathcal{F}_i$ such that $\mathcal{F}_i \subset X^{(i)}$ for every i, $3 \le i \le k$, and for $2 \le r < k$, every r-set not containing any set in $\bigcup_{i=3}^{r} \mathcal{F}_i$ is contained in precisely one element of $\mathcal{F}_{r+1}$. Show that

$$|\mathcal{F}_r| = n(n-1)(n-3)\ldots(n-2^{r-2}+1)/r!$$

for all r, $3 \le r \le k$.

Note that $\mathcal{F}_3$ is an STS, so $\mathcal{F}$ is an elaboration on a triple system. Steiner (1853) asked whether such a system could be constructed for all k and n such that

$$r!|n(n-1)(n-3)\ldots(n-2^{r-2}+1), \quad r = 3, \ldots, k.$$

§15. WEAKLY SATURATED HYPERGRAPHS

Recall that an r-graph $G = (X, \mathcal{E})$ is $(r+s)$-saturated if (i) it does not contain a complete r-graph of order $r+s$ and (ii) if we add to it any of the missing hyperedges (r-sets) then the new graph will contain a $K_{r+s}^{(r)}$.

In §9 we determined the minimal number of edges in an $(r+s)$-saturated r-graph of order n. It turned out that the second of the two conditions was sufficient to ensure that the r-graph has quite a few edges. Having proved the main result by a circuitous route, we noticed that the following special case of a theorem of Bollobás (1965) implies the result rather easily: if the system $\{A_i, B_i : A_i \in X^{(a)}, B_i \in X^{(b)}, i \in I\}$ is such that $A_i \cap B_j = \emptyset$ iff $i = j$ then

$$|I| \leq \binom{a+b}{a}. \tag{1}$$

To define a weakly saturated hypergraph we relax the condition in the definition of a saturated hypergraph. An r-graph $G = (X, \mathcal{E})$ is *weakly $(r+s)$-saturated* if there is a nested sequence of r-graphs $G = G_0 \subset G_1 \subset \ldots \subset G_l$ such that G_{i+1} has one more edge than G_i, G_{i+1} has at least one more $K_{r+s}^{(r)}$ than G_i and $G_l = (X, X^{(r)})$, i.e. the sequence ends with the complete graph. How few edges can G_0 have? (Usually, one also asks that $G = G_0$ has no $K_{r+s}^{(r)}$, but as we are interested in the minimal number of edges of G, this condition can be omitted.) Thus G is weakly $(r+s)$-saturated if, starting with G, we can add to our graph edges one by one in such a way that each new edge is the last missing edge of a $K_{r+s}^{(r)}$ and eventually we can end up with the complete graph $K_n^{(r)}$.

Clearly every $(r+s)$-saturated r-graph is weakly $(r+s)$-saturated so a weakly $(r+s)$-saturated graph need not have more than $\binom{n}{r} - \binom{n-s}{r}$

edges. It was proved by Frankl (1982), Kalai (1985) and Alon (1986) that, in fact, a weakly $(r+s)$-saturated r-graph of order n must have at least $\binom{n}{r} - \binom{n-s}{r}$ edges. This is a substantial extension of Theorem 9.1, because though there is only one $(r+s)$-saturated r-graph with this many edges, there are plenty of extremal hypergraphs for the problem about weakly saturated hypergraphs. Even in the simplest case, namely when $r = 2$ and $s = 1$, we have a rather large set of extremal graphs: the set of trees of order n. (Ex. 2).

The main reason for presenting this extension of Theorem 9.1 is the beauty of the proof given by Frankl, Kalai and Alon: the proof is not only elegant but it also uses a very surprising tool. In fact, having decided on the tool, the rest is just routine. What then is this surprising tool? The exterior algebra! Having revealed this secret, it would be possible to belittle the achievement by saying that this tool is only to be expected since $\binom{n}{r}$ and $\binom{n-s}{r}$ are clearly dimensions of exterior powers. However, this attitude would be most unfair, for the interaction between hypergraphs and exterior algebras is an unexpected phenomenon. As we shall see, the theorem about weakly saturated hypergraphs is tailor-made for a proof by exterior powers.

It happens to be rather easy to express the size of an r-graph in terms of exterior powers, but to make use of this expression is a rather different matter.

Let V be the Euclidean space $\Re^n$ with canonical basis $e_1, e_2, \dots, e_n$ and let $\Lambda V = \sum_{r=0}^{n} \Lambda^r V$ be the *exterior algebra* on V. Thus $\Lambda^r V$ has a basis of the form $\{e_{i_1} \wedge \dots \wedge e_{i_r} : 1 \le i_1 < \dots < i_r \le n\}$, $\dim \Lambda^r V = \binom{n}{r}$ and $\dim \Lambda V = \sum_{r=0}^{n} \binom{n}{r} = 2^r$. Endow ΛV with the inner product induced by that of V, i.e. let the above basis of $\Lambda^r V$ be an orthonormal basis and for $r \ne s$ let $\Lambda^r V$ be orthogonal to $\Lambda^s V$. Write (f, g) for the inner product of f and g in ΛV.

There is a natural way of identifying an r-graph $G = (X, \mathcal{E})$ with the subspace $W(G)$ of $\Lambda^r V$:

$$W(G) = \operatorname{lin}\{e_{i_1} \wedge \dots \wedge e_{i_r} : \{i_1, \dots, i_r\} \in \mathcal{E}\}.$$

By definition, $e(G) = \dim W(G)$. How can we estimate this quantity?

Let $(f_i)_1^n$ be an orthonormal basis of V in a general position with respect to $(e_i)_1^n$. Let Z be the linear span of n vectors of the form $f_{i_1} \wedge \dots \wedge f_{i_r}$. Then, since the f_i's are in a general position, $W(G) \cap Z = \{0\}$ iff $e(G) \le \binom{n}{r} - m$ and $W(G) + Z = \Lambda^r V$ iff $e(G) \ge \binom{n}{r} - m$. Therefore, in theory, to determine $e(G)$ all we have to do is look at the spaces $W(G) \cap Z$ and $W(G) + Z$ and decide for what choice of m they are $\{0\}$ and $\Lambda^r V$, respectively. Needless to say, it would be unreasonable to expect this to be a universally applicable method. However, as we shall show now, this is an ideal method for giving a lower bound for $e(G)$ when G is weakly K^{r+s}-saturated.

Before stating the key lemma, we introduce some additional notation. For a set $A = \{i_1, \ldots, i_r\} \subset X = [n]$, $i_1 < \ldots < i_r$, define $e_A = e_{i_1} \wedge \ldots \wedge e_{i_r}$ and $f_A = f_{i_1} \wedge \ldots \wedge f_{i_r}$. Given $f, g \in \Lambda V$, the map $h \to (h \wedge f, g)$ is a linear functional on ΛV so there is a unique vector $f * g$ representing it, i.e. satisfying

$$(h \wedge f, g) = (h, f * g) \quad \text{for all } h \in \Lambda V.$$

In a vague sense, $f * g$ is g divided by f. Indeed, for a fixed f the map $g \to f * g$ is a linear map; if $A \subset X$ consists of the last k elements of $B \subset X$ then $f_A * f_B = f_{B \setminus A}$, also if $A \subset B \subset X$ then $f_A * f_B$ is either $f_{B\setminus A}$ or $-f_{B\setminus A}$; if $A \not\subset B$ then $f_A * f_B = 0$. In particular, if $m = \sum\{c_A f_A : A \in X^{(r)}\}$ for some coefficients $c_A \in \Re$ then $f_i * m$ iff $i \in A$ implies $c_A = 0$.

Define

$$Z_s^{(r)} = \{m \in \Lambda^r V : f_1 * m = f_2 * m = \ldots = f_s * m = 0\}.$$

Then clearly

$$Z_s^{(r)} = f_S * \Lambda^{r+s} V \subset \Lambda^r V$$

where S=[s] and $\dim Z_s^{(r)}$ is the number of r-sets in X containing no element of $\{1, 2, \ldots, s\}$ so $\dim Z_s^{(r)} = \binom{n-s}{r}$.

Let G_1 be the complete r-graph with vertex set $T = [r+s]$ and write G_0 for the r-graph obtained from G_1 by deleting the hyperedge $A_1 = \{s+1, s+2, \ldots, s+r\} = T \setminus S$.

Lemma 1. $Z_s^{(r)} + W(G_0) = Z_s^{(r)} + W(G_1)$.

Proof. Let $z = f_S * e_T$. Then $z \in Z_s^{(r)}$ and $z = \sum_{A \in T^{(r)}} c_A e_A$, where $c_A \in \Re$. Since $(e_i)_1^n$ and $(f_i)_1^n$ are in general position, all coefficients c_A are non-zero. (This is precisely what we need of the two orthonormal bases). Hence

$$e_{A_1} = c_{A_1}^{-1}\{z - \sum_{\substack{A \in T^{(r)} \\ A \neq A_1}} c_A e_A\} \in Z_s^{(r)} + W(G_0)$$

so

$$\begin{aligned} Z_s^{(r)} + W(G_1) &= Z_s^{(r)} + W\{W(G_0) + \text{lin}\{e_{A_1}\}\} \\ &\subset Z_s^{(r)} + W(G_0) + Z_s^{(r)} + W(G_0) \\ &= Z_s^{(r)} + W(G_0) \end{aligned}$$ ■

Theorem 2. *Let G be a weakly $(r+s)$-saturated r-graph of order n. Then $e(G) \geq \binom{n}{r} - \binom{n-s}{r}$.*

Proof. As usual let us take G with vertex set $X = [n]$ and let V, $(e_i)_1^n$, $(f_i)_1^n$ and $Z_s^{(r)}$ be as before. Let $G_0 = G \subset G_1 \subset \ldots \subset G_m$ be a sequence of r-graphs showing that G is weakly $(r+s)$-saturated, i.e. let this sequence be such that G_m is the complete r-graph on X, $e(G_i) = e(G_{i-1}) + 1$, and the only hyperedge of G_i which is not in G_{i-1} is contained in a complete subgraph of G_i of order $r+s$. Then by Lemma 1 we have

$$Z_s^{(r)} + W(G_0) = Z_s^{(r)} + W(G_1) = \ldots = Z_s^{(r)} + W(G_m).$$

As $W(G_m) = \Lambda^{(r)}V$, this gives

$$Z_s^{(r)} + W(G_0) = \Lambda^{(r)}V$$

so

$$\begin{aligned} e(G) = e(G_0) = \dim W(G_0) &\geq \dim \Lambda^{(r)}V - \dim Z_s^{(r)} \\ &= \binom{n}{r} - \binom{n-s}{r}. \end{aligned}$$ ■

In §9 we pointed out how a special case of Theorem 9.2′, restated in this section as inequality (1), implied Theorem 9.1. The simple argument given in §9 is easily reversed to give the following extension of (1).

Corollary 3. *Let $A_i \in X^{(a)}$, $B_i \in X^{(b)}$, $i = 1, \ldots, m$, be such that $A_i \cap B_i = \emptyset$ for all i and $A_i \cap B_j \neq \emptyset$ if $1 \leq j < i \leq m$. Then $m \leq \binom{a+b}{a}$.*

Proof. Let $\mathcal{E} = X^{(a)} \setminus \{A_1, \ldots, A_m\}$ and $C_i = X \setminus B_i$, $i = 1, \ldots, m$. We claim that the a-uniform hypergraph $G = (X, \mathcal{E})$ is weakly c-saturated where $c = |C_i| = n - b$. Indeed, let $\mathcal{E}_i = \mathcal{E} \cup \{A_1, \ldots, A_i\}$ and $G_i = (X, \mathcal{E}_i)$. Then $C_i^{(a)} \subset \mathcal{E}_i$ because if $i < j$ then $A_j \cap B_i \neq \emptyset$.

Hence A_i, the edge turning G_{i-1} into G_i, completes a $K_c^{(a)}$, namely $C_i^{(a)}$.

As G is weakly c-saturated, $e(G) = \binom{n}{a} - m \geq \binom{n}{a} - \binom{n-c+a}{a} = \binom{n}{a} - \binom{a+b}{a}$ so $m \geq \binom{a+b}{a}$. ∎

Kalai (1984) used a beautiful connection between weakly saturated graphs and rigid graphs to prove Theorem 2 for $r = 2$. Given a graph $G = (V, E)$ with $V = [n]$, an *s-embedding* of G is a function $\mathbf{p} : V \to \Re^s$, say $\mathbf{p}(i) = p_i$, $i = 1, \ldots, n$. We call $\mathbf{p}$ *rigid* if whenever $p_i : \Re \to \Re^s$ are continuous functions such that $p_i(0) = p_i$ and $|p_i(t) - p_j(t)| = |p_i - p_j|$ whenever $ij \in E$, then we also have $|p_i(t) - p_j(t)| = |p_i - p_j|$ for all $i, j \in V$ and $t \geq 0$ (and so the map $p_i \to p_i(t)$ extends to an isometry of $\Re^s$). Thus $\mathbf{p}$ is rigid if every continuous deformation of it preserving the distance between any two adjacent vertices also preserves the distance between any two non-adjacent vertices. An embedding which is not rigid is called *flexible*.

Asimow and Roth (1978) proved that every graph of order n and size at most $sn - \binom{s+1}{2} - 1$ is such that almost every embedding $\mathbf{p} \in \Re^s + \Re^s + \ldots + \Re^s = \Re^{sn}$ is flexible. Intuitively, this follows from the fact that there are sn degrees of freedom to move the vertices and the dimension of the group of affine rigid motions of $\Re^s$ is $\binom{s+1}{2}$. As Kalai (1984) pointed out, if G is weakly $(s+2)$-saturated and $\mathbf{p}$ is an s-embedding of it in which the points $p_1, \ldots, p_n$ are in general position then $\mathbf{p}$ is rigid (Ex. 3). This follows from the fact that in $\Re^s$ we cannot move a point continuously if we want to preserve its distance from each of $s+1$ points in general position. Putting these two facts together, we see that an $(s+2)$-saturated graph of order n has at least $sn - \binom{s+1}{2} = \binom{n}{2} - \binom{n-s}{2}$ edges.

Exercises

1. Let $\mathcal{H}$ be a collection of finite r-graphs. For an r-graph G, let $h(G)$ be the number of subgraphs isomorphic to some member of $\mathcal{H}$. Call G *$\mathcal{H}$-closed* if $h(G) = h(G^+)$ whenever G^+ is obtained from G by adding a single hyperedge. Show that for every G there is a unique minimal closed r-graph $\overline{G}$ containing G; call $\overline{G}$ the *$\mathcal{H}$-closure* of G. Show that G is weakly $(r+s)$-saturated iff its $\{K_{r+s}^{(r)}\}$-closure is $K_n^{(r)}$.

2. Show that every weakly 3-saturated (edge) graph of order $n \geq 3$ has at least $n-1$ edges. Show also that a graph of order $n \geq 3$ and size $n-1$ is weakly 3-saturated iff it is a tree.

3. Let $G = (V, E)$ be an $(s+2)$-saturated graph with $V = [n]$ and let $\mathbf{p} : V \to \mathfrak{R}^s$ be an embedding such that $p_1, \ldots, p_n \in \mathfrak{R}^s$ are in general position. Prove that the embedding $\mathbf{p}$ is rigid. (Kalai (1984))

4. Let us say that a set system $\mathcal{A}$ is a *multiple cover of the s-sets it contains* if whenever $E \subset A \in \mathcal{A}$ and $E \in X^{(s)}$ then there is a set $B \in \mathcal{A}$, $B \neq A$, such that $E \subset B$.

Note that $\mathcal{F} = X_x^{(r)} = \{F \in X^{(r)} : x \in F\}$ is such that no $\mathcal{F}' \subset \mathcal{F}$, $\mathcal{F}' \neq \emptyset$, is a multiple cover of the $(r-1)$-sets it contains.

Peter Frankl used linear algebra, in a way reminiscent of the proofs in this section, to prove that no r-graph with more edges has this property. Here are the steps in the proof.

Let $\mathcal{F} \subset X^{(r)}$, $|\mathcal{F}| > \binom{n-1}{r-1}$. Let U be the vector space over the field $\mathbf{F}_2$ of order 2 with basis $\{e_F : F \in X^{(r)}\}$ and set $U_\mathcal{F} = \mathrm{lin}\{e_F : F \in \mathcal{F}\}$. Let $x \in X$ and $\mathcal{H} = X_x^{(r+1)}$ and define $v_H = \sum\{e_F : F \in H^{(r)}\}$ and $V = \mathrm{lin}\{v_H : H \in \mathcal{H}\}$, for $H \in \mathcal{H}$. Check that $\dim U = \binom{n}{r}$, $\dim U_\mathcal{F} > \binom{n-1}{r-1}$ and $\dim V = \binom{n-1}{r}$ so there is a vector $w \neq 0$,

$$w = \sum\{e_F : F \in \mathcal{F}'\} = \sum\{v_H : H \in \mathcal{H}'\} \in U_\mathcal{F} \cap V.$$

Note that for $H \in \mathcal{H}$ and $F \in X^{(r)}$ either no set in $H^{(r)}$ contains F or else two sets in $H^{(r)}$ contain F. Deduce that $\mathcal{F}' \subset \mathcal{F}$ is a multiple cover of the $(r-1)$-sets it contains.

§16. ISOPERIMETRIC PROBLEMS

In §9 we studied the shadows of set systems and proved the Kruskal-Katona theorem determining the minimal cardinality of the family of all $(r-1)$-sets contained in some set system consisting of r-sets. We also noted that the result can be restated in terms of $(r+1)$-sets containing some r-sets. What happens then if we want to minimize the number of sets just contained in or just containing some of m given sets? Here "B just contains A" means that $A \subset B$ and $|B| \le |A|+1$. Thus B just contains A if $A = B$ or $B \supset A$ and $|A \triangle B| = 1$. Our problem is then the following. For $\mathcal{A} \subset \mathcal{P}(X)$ let $\mathcal{A}_{(1)} = \{B \subset X : B \in \mathcal{A}$ or $|A \triangle B| = 1$ for some $A \in \mathcal{A}\}$. What is $\min\{|\mathcal{A}_{(1)}| : |\mathcal{A}| = m\}$? Equivalently, what is $\min\{|b(\mathcal{A})| : |\mathcal{A}| = m\}$ where $b(\mathcal{A}) = \mathcal{A}_{(1)} \setminus \mathcal{A}$ is the collection of sets in the *boundary* of $\mathcal{A}$?

A somewhat similar, and perhaps even more natural, problem concerns the number of pairs (A, B), where $|A \triangle B| = 1$, A belongs to some set system $\mathcal{A} \subset \mathcal{P}(X)$ with m elements and $B \in \mathcal{P}(X) \setminus \mathcal{A}$. Note that to minimize the number of these pairs the last choice we want is to take $\mathcal{A} \subset X^{(r)}$ for some r, because for any $A \subset X$ precisely n sets $B \subset X$ satisfy $|A \triangle B| = 1$ and if $A \in \mathcal{A} \subset X^{(r)}$ then none of these n sets belong to $\mathcal{A}$. In this case we have mn pairs which is the maximum rather than the minimum.

The appearance of the condition $|A \triangle B| = 1$ indicates that the natural setting for the problems above is the covering graph of the partially ordered set, namely the n-dimensional cube Q^n. Recall that the vertex set of Q^n can be taken to be the set of all 2^n sequences of length n consisting of 0's and 1's and two sequences are joined iff they differ in precisely one place. Equivalently, and this is the form we shall use, $V(Q^n) = \{0, 1, 2, \dots, 2^n - 1\}$ and vertex i is joined to vertex j iff the binary expansions of i and j differ in just one digit. The problems above are special cases of isoperimetric inequalities in graphs.

What is an isoperimetric inequality in a graph? An inequality, preferably a best possible inequality, relating the order of a subgraph to the size of its boundary. There are several possibilities for defining the boundary of a subgraph, with two of those possibilities being rather prominent. For an induced subgraph H of a graph G, the *edge boundary* of H is the set of edges joining a vertex in H to a vertex not in H. We write $b_e(H;G)$ for the number of edges in the edge boundary. Similarly, the *vertex boundary* of H is the set of vertices not in H joined to some vertices in H; the number of vertices in the vertex boundary is denoted by $b_v(H;G)$.

An *isoperimetric inequality* is a lower bound for the size of the boundary in terms of the order or size of the subgraph. Thus one of the best isoperimetric inequalities for the edge boundary is an explicit expression for

$$b_e(m;G) = \min\{b_e(H;G) : H \subset G, |H| = m\}$$

and one of the best isoperimetric inequalities for the vertex boundary is an explicit expression for

$$b_v(m;G) = \min\{b_v(H;G) : H \subset G, |H| = m\}.$$

The main aim of this section is to determine $b_e(m;Q^n)$ and $b_v(m;Q^n)$ and thereby to answer the two questions we posed about sets systems.

The function $b_e(m;Q^n)$ was determined by Hart (1976); related results had been proved earlier by Harper (1964, 1966, 1967) and Bernstein (1967). The function $b_e(m;Q^n)$ plays an important role in the study of random subgraphs of the cube (see Bollobás (1985, Ch. XIV)).

Which subgraph H of order m in Q^n has a small edge boundary? One that spans many edges. If $m = 2^k$ then the best we can do is to take a k-dimensional subcube of Q^n. If $2^k \leq m < 2^{(k+1)}$ then we take one half (say, the bottom face) of a $(k+1)$-dimensional cube, and take $m - 2^k$ more vertices in the other half (say, the top face) as best we can. For example, if $m = 2^k + 2^l$ for some $0 \leq l < k < n$ then the best choice is the bottom face of a $(k+1)$-dimensional cube together with an l-dimensional subcube of the top face.

Thus the extremal function seems rather sweet and simple and we may expect to run into some difficulties only in proving it. Curiously enough, the situation is rather different. The proof will be immediate, provided we can show that the very innocent-looking extremal function satisfies a certain very simple inequality (see (1) below). But the proof of the inequality is not as simple as one would expect. Note that this state of affairs is in marked contrast with most of the extremal problems: usually the functions are rather simple and the difficulties in the proofs have very little to do with the actual form of the functions. For example, one would have to come up with a very peculiar inequality indeed to make it difficult to prove for $sn - \binom{s+1}{2}$, the extremal function for the graph case in Theorem 15.2.

Let us get down to some mathematics. Since the cube Q^n is n-regular, every induced subgraph $H \subset Q^n$ of order m satisfies

$$b_e(H) = mn - 2e(H),$$

where $e(H)$ is, as usual, the number of edges in H. Therefore

$$b_e(m; Q^n) = mn - 2e(m; Q^n)$$

where

$$e(m; Q^n) = \max\{e(H) : H \subset Q^n, |H| = m\}.$$

Denote by $h(i)$ the sum of digits in the binary expansion of i, and for $0 \leq l < m$ set

$$f(l, m) = \sum_{l \leq i < m} h(i).$$

Lemma 1. *If* $1 \leq k \leq l$ *then*

$$f(l, l+k) \geq f(0, k) + k. \tag{1}$$

Proof. What can we say about the function $f(l, l+k)$? First of all,

$$h(i) + h(2^r - 1 - i) = r$$

for all i, $0 \leq i \leq 2^r - 1$, because for $1 \leq j \leq r$ the jth digit of i is 1 iff the jth digit of $2^r - 1 - i$ is 0. Consequently

$$f(l, l+k) + f(2^r - l - k, 2^r - l) = rk \tag{2}$$

whenever $l + k \leq 2^r$.

Furthermore, for $j \geq 1$, in the binary expansions of the numbers $0, 1, 2, \ldots$ the jth digit is 0 2^j times, then 1 2^j times, then 0 2^j times, then 1 2^j times, etc. Therefore the sum of the jth digits of k consecutive numbers is minimal if the first block of 0's is as long as possible. This shows that

$$f(l, l+k) \geq f(0, k) \tag{3}$$

for all $l \geq 0$ and

$$f(l, l+k) \geq f(2^r, 2^r+k) = f(0, k) + k \tag{4}$$

whenever $k \leq 2^r \leq l$.

Now let us turn to the proof of (1). We shall apply induction on k. As the inequality is trivial for $k = 1$, we turn to the induction step. Let $k \geq 2$ and suppose that inequality (1) holds for smaller values of k. Define $r \geq 1$ by $2^{r-1} \leq k < 2^r$. We may assume that $2^{r-1} < l < 2^r$ since otherwise (1) is implied by inequality (4). Then $l < 2^r < l + k$ so by (2), (4) and the induction hypothesis we have

$$\begin{aligned} f(l, l+k) &= f(l, 2^r) + f(2^r, l+k) \\ &= (2^r - l)r - f(0, 2^r - l) + f(2^r, l+k) \\ &\geq (2^r - l)r - f(0, 2^r - l) + f(0, l+k-2^r) + l + k - 2^r \\ &\geq (2^r - l)r - f(2^r - k, 2^r - k + 2^r - l) + 2^r - l \\ &\qquad\qquad + f(0, l+k-2^r) + l + k - 2^r \\ &= f(l+k-2^r, k) + f(0, l+k-2^r) + k = f(0, k) + k. \end{aligned}$$

The penultimate equality followed from (2) since $2^{r+1} - k - l \leq 2^{r+1} - 2^r = 2^r$. This completes the proof of the induction step. ∎

Note that so far we haven't even mentioned the function $b_e(H; Q^n)$ let alone tried to determine its minimum. All we have done is proved a rather simple inequality, (1), about the function $f(l, l+k)$, the sum of the binary digits of $l, l+1, \ldots, l+k-1$. Nevertheless, the proof of the following result of Harper, Bernstein and Hart is just around the corner.

Theorem 2. *For $1 \leq m \leq 2^n$ we have*

$$e(m; Q^n) = f(0, m) \qquad \textit{and} \qquad b_e(m; Q^n) = mn - 2f(0, m).$$

Proof. The first inequality implies the second. Furthermore, the set $W = \{0, 1, \ldots, m-1\}$ spans precisely $f(0, m)$ edges because the vertex i is joined to exactly $h(i)$ vertices j with $j < i$. Hence $e(m; Q^n) \geq f(0, m)$

so all we have to prove is that $e(m; Q^n) \le f(0, m)$. We shall prove this by induction on n.

The case $n = 1$ is trivial so let us turn to the induction step. Split Q^n into two $(n-1)$-dimensional cubes, the top face and the bottom face, say. Let W be an m-subset of $V(Q^n) = \{0, 1, \ldots, 2^n - 1\}$ having m_1 vertices on the top face and m_2 vertices on the bottom face. Say $m_1 \le m_2$. Every vertex of the bottom face is joined to precisely one vertex of the top face and every vertex of the top face is joined to precisely one vertex of the bottom face. Hence by the induction hypothesis, the number of edges spanned by W is at most

$$f(0, m_1) + f(0, m_2) + m_1 = f(0, m) - f(m_2, m_1 + m_2) + f(0, m_1) + m_1.$$

By Lemma 1 this expression is at most $f(0, m)$, as required. ∎

Let us turn to the first problem mentioned in this section. For what system of m subsets of X is the collection of subsets within distance 1 of the system smallest? In other words, how should we choose m vertices of Q^n to make the set of vertices within distance 1 smallest? If $m = \sum_{i=0}^{k} \binom{n}{i}$ for some k, then for any vertex $x \in V(Q^n)$ there are precisely m vertices in the ball $B(x, k) = \{y \in V(Q^n) : d(x, y) \le k\}$ of centre x and radius k and there are $\binom{n}{k+1}$ vertices in the vertex boundary. Thus $b_v(m; Q^n) \le \binom{n}{k+1}$. Harper (1966) was the first to prove that this is the best we can do; simpler proofs were given by Katona (1975) and Frankl and Füredi (1981). Here we shall present the proof of Frankl and Füredi.

Call a set system $\mathcal{A} \subset \mathcal{P}(X)$ a *Hamming ball with centre* $C \in \mathcal{P}(X)$ if

$$\mathcal{B}(C, r) \subset \mathcal{A} \subset \mathcal{B}(C, r+1)$$

for some non-negative integer r, where $\mathcal{B}(C, r) = \{A \subset X : |A \Delta C| \le r\}$ is the ball of radius r and centre C in $\mathcal{P}(X)$. For a set system $\mathcal{A} \subset \mathcal{P}(X)$, define $\mathcal{A}_{(d)}$ as the collection of sets at distance at most d from $\mathcal{A}$.

$$\mathcal{A}_{(d)} = \{B \subset X : |A \Delta B| \le d \text{ for some } A \in \mathcal{A}\}.$$

Note that $\mathcal{B}(C, r) = \{C\}_{(r)}$. Instead of bounding $|\mathcal{A}_{(d)}|$ from below, we shall put an upper bound on the number of sets far from $\mathcal{A}$, i.e. on the number of sets in

$$\mathcal{P}(X) \setminus \mathcal{A}_{(d)} = \{B \subset X : |A \Delta B| \ge d + 1 \text{ for all } A \in \mathcal{A}\}.$$

For $\mathcal{A}, \mathcal{B} \subset \mathcal{P}(X)$ define the *distance* between $\mathcal{A}$ and $\mathcal{B}$ as $d(\mathcal{A}, \mathcal{B}) = \min\{d(A, B) = |A \Delta B| : A \in \mathcal{A}, B \in \mathcal{B}\}$.

Theorem 3. *Let $\mathcal{A}$ and $\mathcal{B}$ be non-empty set systems. Then we can find a Hamming ball $\mathcal{A}_0$ with centre X and a Hamming ball $\mathcal{B}_0$ with centre $\emptyset$ such that $|\mathcal{A}_0| = |\mathcal{A}|$, $|\mathcal{B}_0| = |\mathcal{B}|$ and $d(\mathcal{A}_0, \mathcal{B}_0) \geq d(\mathcal{A}, \mathcal{B})$.*

Proof. Let $a = |\mathcal{A}|$, $b = |\mathcal{B}|$, $d = d(\mathcal{A}, \mathcal{B})$ and let $\mathcal{A}_0, \mathcal{B}_0 \subset \mathcal{P}(X)$ be such that $|\mathcal{A}_0| = a$, $|\mathcal{B}_0| = b$, $d(\mathcal{A}_0, \mathcal{B}_0) \geq d$ and

$$\sigma(\mathcal{A}_0, \mathcal{B}_0) = \sum_{A \in \mathcal{A}_0} |A| - \sum_{B \in \mathcal{B}_0} |B| \tag{6}$$

is maximal. We claim that $\mathcal{A}_0$ and $\mathcal{B}_0$ are appropriate Hamming balls.

It is immediate though, as it happens, not too important that $\mathcal{A}_0$ is monotone increasing and $\mathcal{B}_0$ is monotone decreasing. Indeed, if $h \in X$ then by adding h to every set in $\mathcal{A}_0$ to which it can be added (i.e. replacing $\mathcal{A}_0$ by $\tilde{T}_h(\mathcal{A}_0)$ for the operation $\tilde{T}_h$ in §13) and taking away h from every set in $\mathcal{B}_0$ from which it can be taken away, we find new systems at distance at least d apart. Since (6) must not decrease, these operations do not change $\mathcal{A}_0$ and $\mathcal{B}_0$ for any h so $\mathcal{A}_0$ is monotone increasing and $\mathcal{B}_0$ is monotone decreasing.

Suppose it is not true that $\mathcal{A}_0$ is an X-centred and $\mathcal{B}_0$ is an $\emptyset$-centred Hamming ball. Then, by symmetry, we may assume that there are $A_0 \in \mathcal{A}_0$ and $A_0^* \notin \mathcal{A}_0$ such that $|A_0| < |A_0^*|$ and

$$|A_0 \triangle A_0^*| \leq |C \triangle C^*|$$

if $C \in \mathcal{A}_0$, $C^* \notin \mathcal{A}_0$ and $|C| < |C^*|$. Then, in particular, the system $\mathcal{F}_0 = \{F \subset X : |F| = |A_0^*|, |A_0 \triangle F| < |A_0 \triangle A_0^*|\}$ is contained in $\mathcal{A}_0$. Set $U = A_0 \setminus A_0^*$ and $V = A_0^* \setminus A_0$ so that $|U| < |V|$. Note that $U \neq \emptyset$ because all sets containing A_0 belong to $\mathcal{A}_0$.

Let us use the sets U and V to transform the systems $\mathcal{A}_0$ and $\mathcal{B}_0$ and thereby arrive at a contradiction. For $A \in \mathcal{A}_0$ set

$$U(A) = \begin{cases} A \cup V \setminus U & \text{if } U \subset A \subset X \setminus V \text{ and } A \cup V \setminus U \notin \mathcal{A}_0 \\ A & \text{otherwise,} \end{cases}$$

and, analogously, for $B \in \mathcal{B}_0$ set

$$D(B) = \begin{cases} B \cup U \setminus V & \text{if } V \subset B \subset X \setminus U \text{ and } B \cup U \setminus V \notin \mathcal{B}_0 \\ B & \text{otherwise,} \end{cases}$$

Define $\mathcal{A}_1 = \{U(A) : A \in \mathcal{A}_0\}$ and $\mathcal{B}_1 = \{D(B) : B \in \mathcal{B}_0\}$. Thus $\mathcal{A}_1$ was obtained by moving the sets in $\mathcal{A}_0$ "up", replacing U by the bigger set

V and $\mathcal{B}_1$ was obtained by moving the sets in $\mathcal{B}_0$ "down", replacing V by the smaller set U.

Clearly $|\mathcal{A}_1| = |\mathcal{A}_0|$, $|\mathcal{B}_1| = |\mathcal{B}_0|$ and $\sigma(\mathcal{A}_1, \mathcal{B}_1) > \sigma(\mathcal{A}_0, \mathcal{B}_0)$. Hence, to arrive at a contradiction and so complete the proof, all we have to check is that $d(\mathcal{A}_1, \mathcal{B}_1) \geq d$, that is $d(A, B) \geq d$ for all $A \in \mathcal{A}_1$, $B \in \mathcal{B}_1$.

If $A \in \mathcal{A}_1 \cap \mathcal{A}_0$ and $B \in \mathcal{B}_1 \cap \mathcal{B}_0$ then, trivially, $d(A, B) \geq d$.

If $A' \in \mathcal{A}_1 \setminus \mathcal{A}_0$, $B' \in \mathcal{B}_1 \setminus \mathcal{B}_0$, say $A' = A \cup V \setminus U$ and $B' = B \cup U \setminus V$, then $d(A', B') = d(A, B) \geq d$.

The cases $A' \in \mathcal{A}_1 \setminus \mathcal{A}_0$, $B \in \mathcal{B}_1 \cap \mathcal{B}_0$ and $A \in \mathcal{A}_1 \cap \mathcal{A}_0$, $B' \in \mathcal{B}_1 \setminus \mathcal{B}_0$ are similar, so let us assume that $A' \in \mathcal{A}_1 \setminus \mathcal{A}_0$ and $B \in \mathcal{B}_1 \cap \mathcal{B}_0$, say $A' = A \cup V \setminus U$, $A \in \mathcal{A}_0$. If $V \subset B \subset X \setminus U$ then $B' = B \cup U \setminus V \in \mathcal{B}_0$ and so $A' \triangle B = A \triangle B'$, implying $d(A', B) = d(A, B') \geq d(\mathcal{A}_0, \mathcal{B}_0) \geq d$.

Suppose then that $V \subset B \subset X \setminus U$ does not hold. Then there are $u \in U$ and $v \in V$ such that not both of $v \in B$ and $u \in U \setminus B$ hold. Set

$$\overline{A} = A \cup (V \setminus \{v\}) \setminus (U \setminus \{u\}) = A' \cup \{u\} \setminus \{v\}.$$

Then $|\overline{A}| = |A'| > |A|$ and $|A \triangle \overline{A}| < |U| + |V| = |A_0 \triangle A_0^*|$ so the choice of A_0 and A_0^* implies that $\overline{A} \in \mathcal{A}_0$. Finally, by the choice of u and v,

$$\begin{aligned} |A' \triangle B| &= \left|(\overline{A} \cup \{v\} \setminus \{u\}) \triangle B\right| \\ &= |\overline{A} \triangle B| + (-1)^{|\{v\} \cap B|} - (-1)^{|\{u\} \cap B|} \\ &\geq |\overline{A} \triangle B| \geq d(\mathcal{A}_0, \mathcal{B}_0) \geq d. \end{aligned}$$ ■

Theorem 3 implies immediately a complete solution of the isoperimetric problem for the vertex boundary in the cube. Indeed, if $\mathcal{A} \subset \mathcal{P}(X)$, $|\mathcal{A}| = m$ and $\sum_{i=r+1}^{n} \binom{n}{i} < m \leq \sum_{i=r}^{n} \binom{n}{i}$ then there is a Hamming ball $\mathcal{A}_0$ of centre X consisting of m sets whose boundary is at most as large as the boundary of $\mathcal{A}$. Now if $\mathcal{F}_0 = \mathcal{A}_0 \cap X^{(r)}$ then

$$b(\mathcal{A}_0) = X^{(\geq r)} \cup \partial \mathcal{F}_0,$$

where $\partial \mathcal{F}_0 = \partial_l \mathcal{F}_0 = \{E \in X^{(r-1)} : E \subset F \text{ for some } F \in \mathcal{F}_0\}$ is the lower shadow of $\mathcal{F}_0$. We know from the Kruskal-Katona theorem, Theorem 5.3, that to get a smallest shadow we can choose $\mathcal{F}_0$ to be the set of the first $|\mathcal{F}_0|$ r-sets in the colex order. In particular, with the notation of §5, if $|\mathcal{F}_0| = b^{(r)}(m_r, \ldots, m_s)$ then the minimum of $|\partial \mathcal{F}_0|$ is $b^{(r-1)}(m_r, \ldots, m_s)$. Expressing this in detail, we have the following theorem.

Theorem 4. *Every integer m, $1 \leq m \leq 2^n - 1$, has a unique representation in the form*

$$m = \sum_{i=r+1}^{n} \binom{n}{i} + m', \qquad 0 < m' \leq \binom{n}{r},$$

$$m' = \sum_{j=s}^{r} \binom{m_j}{j}, \qquad 1 \le s \le m_s < m_{s+1} < \ldots < m_r.$$

Then

$$b_v(m; Q^n) = \binom{n}{r} - m' + \sum_{j=s}^{r} \binom{m_j}{j-1}.$$

■

As we do not wish to test the patience of the reader, we state a special case of the result above which is, undoubtedly, much more attractive than Theorem 4.

Theorem 5. *Let* $m = \sum_{i=0}^{r} \binom{n}{i}$. *Then*

$$b_v(m; Q^n) = \binom{n}{r+1}.$$

In other words, if $\mathcal{A} \subset \mathcal{P}(X)$ *and* $|\mathcal{A}| = m$ *then*

$$|\mathcal{A}_{(1)}| \ge \sum_{i=0}^{r+1} \binom{n}{i} \qquad \text{and} \qquad |b(\mathcal{A})| \ge \binom{n}{r+1}.$$

Every Hamming ball $B(C, r)$ *shows that these inequalities are best possible.*

■

In §5 we made use of the fact that if $\mathcal{F}_0 \subset X^{(r)}$ is an initial segment of the colex order then so is the lower shadow $\partial_l \mathcal{F}_0$, the lower shadow of that, $\partial_l^2 \mathcal{F}_0$, and so on. Combining this with Theorem 3, we get an appropriate extension of Theorem 4. Let us just state it in its simplest form, which follows directly from Theorem 3.

Theorem 6. *If* $\mathcal{A} \subset \mathcal{P}(X)$ *and* $|\mathcal{A}| = \sum_{i=0}^{r} \binom{n}{i}$ *then for* $1 \le d \le n - r$ *we have*

$$|\mathcal{A}_{(d)}| \ge \sum_{i=0}^{r+d} \binom{n}{i}.$$

Every Hamming ball $B(C, r)$ *shows that this inequality is best possible.*

■

Exercises

1. Deduce from Theorem 2 that for all $1 \le m \le 2^n$ we have

$$e(m; Q^n) \le \frac{m}{2} \lceil \log_2 m \rceil.$$

2. Check that if $\mathcal{A} \subset \mathcal{P}(X)$ and $h \in X$ then

$$\left| b\left(\tilde{T}_h(\mathcal{A})\right) \right| \le |b(\mathcal{A})|,$$

where

$$\tilde{T}_h(\mathcal{A}) = \left\{ A \setminus \{h\} : h \in A \in \mathcal{A} \right\} \cup \left\{ A \in \mathcal{A} : A \setminus \{h\} \in \mathcal{A} \right\}$$

3. Show that if $\mathcal{A}$ is a monotone increasing set system then

$$\left| b\left(\tilde{R}_{ij}(\mathcal{A})\right) \right| \le |b(\mathcal{A})|$$

where

$$\tilde{R}_{ij}(\mathcal{A}) = \{R_{ij}(A) : A \in \mathcal{A}\} \cup \{A : R_{ij}(A) \in \mathcal{A}\}.$$

and

$$R_{ij} = \begin{cases} (A \setminus \{j\}) \cup \{i\} & \text{if } j \in A \text{ and } i \notin A \\ A & \text{otherwise} \end{cases}$$

4. Let $1 \le k \le n$ and $m = k2^{(k-1)}$. Let H be a subgraph of Q^n containing m edges. (H need not be an induced subgraph.) Show that the cube has at least $(n-k)2^k$ edges not in H incident with some edges of H.

§17. THE TRACE OF A SET SYSTEM

Let S be an infinite set and let $\mathcal{F} \subset \mathcal{P}(S)$. Thus both $\mathcal{F}$ and its members may be infinite. Given a set Y, we call $\mathcal{F} \cap Y = \{F \cap Y : F \in \mathcal{F}\}$ the *trace of* $\mathcal{F}$ *on* Y or the *restriction of* $\mathcal{F}$ to Y. For a natural number k, let

$$f_{\mathcal{F}}(k) = \max\{|\mathcal{F} \cap Y| : Y \in S^{(k)}\}.$$

Clearly $0 \leq f_{\mathcal{F}}(k) \leq 2^k$ for every k. Furthermore, if $l < k$ and $f_{\mathcal{F}}(k) = 2^k$ then $f_{\mathcal{F}}(l) = 2^l$. With what speed can $f_{\mathcal{F}}(k)$ tend to infinity as $k \to \infty$? If $\mathcal{F} = S^{(\leq r)}$ then $f_{\mathcal{F}}(k) = \sum_{i=0}^{r} \binom{k}{i}$ for every k and if $\mathcal{F} = \mathcal{P}(S)$, say, then $f_{\mathcal{F}}(k) = 2^k$ for every k. Can the speed be somewhere between the two? Can $f_{\mathcal{F}}(k)$ be subexponential but still faster than any polynomial? This question, posed by Erdős in 1970, was answered by Sauer (1972) and Perles and Shelah (see Shelah (1972)) in the negative: if for every r there is a k such that $f_{\mathcal{F}}(k) > k^r$ then $f_{\mathcal{F}}(k) = 2^k$ for every k. In fact, this is an immediate consequence of a theorem on (finite) families of (finite) sets.

To formulate this theorem, we define the *trace number* of a set system $\mathcal{F} \subset \mathcal{P}(X)$ as

$$\begin{aligned} \operatorname{tr}(\mathcal{F}) &= \max\{m : f_{\mathcal{F}}(m) = 2^m\} \\ &= \max\{|Y| : Y \subset X, \mathcal{F} \cap Y = \mathcal{P}(Y)\}. \end{aligned}$$

Theorem 1. *Suppose* $\mathcal{F} \subset \mathcal{P}(X)$ *and*

$$|\mathcal{F}| > \sum_{i=0}^{k-1} \binom{n}{i}.$$

Then $tr(\mathcal{F}) \geq k$.

Proof. For $x \in X$ set $\mathcal{F}_x = \mathcal{F} \cap (X \setminus \{x\}) = \{A \setminus \{x\} : A \in \mathcal{F}\}$ and let $\varphi_x : \mathcal{F} \to \mathcal{F}_x$ be given by $\varphi_x(A) = A \setminus \{x\}$. Clearly φ_x is a surjection and if $|\varphi_x^{-1}(B)| \geq 2$ then $\varphi_x^{-1}(B) = \{B, B \cup \{x\}\}$ and $x \notin B$. Thus if

$$\mathcal{A}_x = \{A \in \mathcal{F} : x \in A, A \setminus \{x\} \in \mathcal{F}\}$$

and

$$\mathcal{B}_x = \{B \in \mathcal{F} : x \notin B, B \cup \{x\} \in \mathcal{F}\}$$

then

$$|\mathcal{F}| - |\mathcal{F}_x| = |\mathcal{A}_x| = |\mathcal{B}_x| \tag{1}$$

Note also that if $\mathrm{tr}(\mathcal{B}_x) \geq k-1$ then $\mathrm{tr}(\mathcal{F}) \geq k$. Indeed, suppose $\mathcal{B}_x \cap Y = \mathcal{P}(Y)$. Set $Z = Y \cup \{x\}$. Then $\mathcal{F} \cap Z \supset (\mathcal{A}_x \cup \mathcal{B}_x) \cap Z = \mathcal{P}(Z)$ because if $x \in U \subset Z$ then $U - \{x\} = B \cap Y = B \cap Z$ for some $B \in \mathcal{B}_x$ so $U - \{x\} = B \cap Z$ and $U = A \cap Z$, where $A = B \cup \{x\} \in \mathcal{A}_x$.

And now for the actual proof. Let us apply induction on $n+k$. For $n+k=1$ there is nothing to prove. Suppose $n+k \geq 2$ and the result is true for smaller values of $n+k$.

Let $x \in X$. If $|\mathcal{F}_x| > \sum_{j=0}^{k-1} \binom{n-1}{j}$ then $\mathrm{tr}(\mathcal{F}_x) \geq k$ by induction and so $\mathrm{tr}(\mathcal{F}) \geq k$. On the other hand, if $|\mathcal{F}_x| \leq \sum_{j=0}^{k-1} \binom{n-1}{j}$ then, by (1),

$$|\mathcal{B}_x| > \sum_{j=0}^{k-1} \binom{n}{j} - \sum_{j=0}^{k-1} \binom{n-1}{j} = \sum_{j=1}^{k-1} \binom{n-1}{j-1} = \sum_{j=0}^{k-2} \binom{n-1}{j}.$$

Hence, by induction, $\mathrm{tr}(\mathcal{B}_x) \geq k-1$ and so $\mathrm{tr}(\mathcal{F}) \geq k$. ■

Corollary 2. *If $\mathcal{F}$ is a family of subsets of an infinite set S then either $f_{\mathcal{F}}(k) = 2^k$ for every k or else there is an $r \in \mathbb{N}$ such that $f_{\mathcal{F}}(k) \leq k^r$ for every $k \geq 2$.*

Proof. Suppose $f_{\mathcal{F}}(k) \neq 2^k$ for some k. Then, by Theorem 1, $f_{\mathcal{F}}(n) \leq \sum_{i=0}^{k-1} \binom{n}{i} \leq n^k$ for $n > k$. ■

Theorem 1 has been extended in several directions, notably by Karpovsky and Milman (1978), Frankl (1983), Alon (1983) and Alon and Milman (1983). Here we shall confine our attention to the functions $f_{\mathcal{F}}(k)$ for various set systems. Let us introduce the following *arrow notation*: $(m,n) \to (r,s)$ means that $f_{\mathcal{F}}(s) \geq r$ if $\mathcal{F} \subset \mathcal{P}(X)$ and $|\mathcal{F}| \geq m$. In this notation the theorem due to Sauer, Perles and Shelah states that

$$(m,n) \to (2^k, k) \quad \text{if} \quad m > \sum_{i=0}^{k-1} \binom{n}{i}. \tag{2}$$

The reader may well recall that a very simple aspect of the function $f_{\mathcal{F}}(k)$ was discussed in §2. In the arrow notation Theorem 2.1, due to Bondy (1972), claims precisely that $(n,n) \to (n,n-1)$. Another little result was noted by Bollobás (see Lovász (1979, p. 444)):

$$(m,n) \to (m-1,n-1) \quad \text{if} \quad m \le \lceil 3n/2 \rceil. \tag{3}$$

Frankl (1983) showed that in proving an arrow relation $(m,n) \to (r,s)$ we may restrict our attention to ideals, i.e. to monotone decreasing families. Essentially the same result was proved independently by Alon (1983). Using this theorem, Frankl and Alon gave quick proofs of all the results above (see Exercises 1 and 2). The theorem itself is an immediate consequence of a lemma concerning the operator $\tilde{T}_h$ used in §§13 and 16. Recall that for $\mathcal{F} \subset \mathcal{P}(X)$ and $h \in X$ the system $\tilde{T}_h(\mathcal{F})$ is obtained from $\mathcal{F}$ by omitting h from every set $A \in \mathcal{F}$ whenever possible, if the omission does not decrease the total number of sets:

$$\tilde{T}_h(\mathcal{F}) = \{A \setminus \{h\} : A \in \mathcal{F}\} \cup \{A \in \mathcal{F} : A \setminus \{h\} \in \mathcal{F}\}.$$

Thus if $B \subset A \subset X$, $h \notin B$ and $A = B \cup \{h\}$ then $\mathcal{F}$ and $\tilde{T}_h(\mathcal{F})$ contain the same number of sets from $\{A, B\}$ and

$$A \in \tilde{T}_h(\mathcal{F}) \quad \text{implies} \quad B \in \tilde{T}_h(\mathcal{F}).$$

Lemma 3. *Let $\mathcal{F} \subset \mathcal{P}(X)$, $h \in X$ and $\mathcal{H} = \tilde{T}_h(\mathcal{F})$. then*

$$f_{\mathcal{H}}(s) \le f_{\mathcal{F}}(s)$$

for every s, $1 \le s \le n$.

Proof. Let $Y \in X^{(s)}$. If $h \notin Y$ then clearly $\mathcal{F} \cap Y = \mathcal{H} \cap Y$.

Assume then that $h \in Y$. Consider all 2^{s-1} pairs of subsets of Y of the form (U,V) where $U \subset V \subset Y$, $h \notin U$ and $V = U \cup \{h\}$. It suffices to show that for each pair (U,V), the system $\mathcal{F} \cap Y$ contains at least as many of U and V as $\mathcal{H} \cap Y$.

If $V \in \mathcal{H} \cap Y$, say $V = D \cap Y$ for some $D \in \mathcal{H}$, then $h \in D$ so, by the definition of $\mathcal{H}$, both D and $C = D \setminus \{h\}$ belong to $\mathcal{F}$. Hence $\mathcal{F} \cap Y$ contains both $U = C \cap Y$ and $V = D \cap Y$.

If $U \in \mathcal{H} \cap Y$, say $U = C \cap Y$ for some $C \in \mathcal{H}$, then either $C \in \mathcal{F}$ or $D = C \cup \{h\} \in \mathcal{F}$. In the first case $\mathcal{F} \cap Y$ contains $U = C \cap Y$ and in the second it contains $V = D \cap Y$. Hence $\mathcal{F} \cap Y$ contains at least one of U and V. ■

Theorem 4. *Suppose m, n, r and s are such that $f_{\mathcal{H}}(s) \geq r$ for every ideal $\mathcal{H} \subset \mathcal{P}(X)$ with $|\mathcal{H}| \geq m$. Then $(m,n) \to (r,s)$.*

Proof. Let $\mathcal{F} \subset \mathcal{P}(X)$ be such that $f_{\mathcal{F}}(s) \leq r-1$. Set $\mathcal{H} = \tilde{T}_1(\tilde{T}_2(\ldots \tilde{T}_n(\mathcal{F})\ldots))$. Then $\mathcal{H}$ is an ideal, $|\mathcal{H}| = |\mathcal{F}|$ and, by Lemma 3, $f_{\mathcal{H}}(s) \leq f_{\mathcal{F}}(s) \leq r-1$. Therefore $|\mathcal{F}| = |\mathcal{H}| \leq m-1$, proving $(m,n) \to (r,s)$. ■

Note that Theorem 1 is an immediate consequence of Theorem 4. Indeed, if $\mathcal{H} \subset \mathcal{P}(X)$ and $|\mathcal{H}| > \sum_{i=0}^{k-1} \binom{n}{i}$ then $\mathcal{H}$ contains a set Y of size at least k. If $\mathcal{H}$ is an ideal then $\mathcal{P}(Y) \subset \mathcal{H}$ so, *a fortiori*, $\mathcal{H} \cap Y = \mathcal{P}(Y)$. Hence, by Theorem 4, $(m,n) \to (2^k, k)$ whenever $m > \sum_{i=0}^{k-1} \binom{n}{i}$.

Exercises

1. Use Theorem 4 to prove that $(m,n) \to (n, n-1)$ if $m \geq n$. (Bondy (1972))

2. Deduce from Theorem 4 that $(m,n) \to (m-1, n-1)$ if $m \leq \lceil 3n/2 \rceil$. Show also that if $m \geq \lceil 3n/2 \rceil + 1$ then $(m,n) \not\to (m-1, n-1)$. (Bollobás, see Lovász (1979, p. 444))

3. Turán's theorem for triangles (see §8) states that if a graph of order n and size m has no triangles then $m \leq \lfloor n^2/4 \rfloor$. Use this result to prove that $(m,n) \to (7,3)$ if $m \geq \lfloor n^2/4 \rfloor + n + 2$. Show also that Turán's theorem for triangles is an immediate consequence of this latter assertion. (Frankl (1983) and Alon (1983))

4. Given natural numbers $p_1, \ldots, p_n$, let $\mathcal{F} = \mathcal{F}(p_1, \ldots, p_n)$ be the set of all functions $f : [n] \to \mathbf{N}$ satisfying $f(i) \leq p_i$ for $1 \leq i \leq n$. Call a family $\mathcal{H} \subset \mathcal{F}$ *monotone* if $f \in \mathcal{H}$ whenever $f \in \mathcal{F}$ and f is dominated by some function in $\mathcal{H}$. Finally, given $\mathcal{G} \subset \mathcal{F}$ and $A \subset X = [n]$ let $\mathcal{G}|A = \{g|A : g \in \mathcal{G}\}$.

Suppose $\mathcal{A} \subset \mathcal{P}(X)$ and the family $\mathcal{G} \subset \mathcal{F}$ is such that $\mathcal{G}|A \neq \mathcal{F}|A$ for all $A \in \mathcal{A}$. Imitate the proof of Theorem 3 to show that there is a monotone family $\mathcal{H}$ such that $|\mathcal{H}| = |\mathcal{G}|$ and $\mathcal{H}|A \neq \mathcal{G}|A$ for all $A \in \mathcal{A}$. (Alon (1983))

§18. PARTITIONING SETS OF VECTORS

In §2 we encountered Hall's fundamental theorem: a finite family $\mathcal{A} = \{A_i : i \in I\}$ of finite sets has a transversal iff $\left|\bigcup_{j\in J} A_j\right| \geq |J|$ for every $J \subset I$ (Theorem 2.2). The essential part of this result is that the trivial necessary condition for the existence of a transversal is also sufficient. Can we characterize those families $\mathcal{A}$ which are unions of k families, each of which has a transversal? Suppose $I = \bigcup_{i=1}^{k} I_i$ and each family $\mathcal{A}_i = \{A_j : j \in I_i\}$ has a transversal. For $J \subset I$ set $J_i = J \cap I_i$ so that $J = \bigcup_{i=1}^{k} J_i$. Then $|J_l| \geq |J|/k$ for some l, so

$$\left|\bigcup_{j\in J} A_j\right| \geq \left|\bigcup_{j\in J_l} A_j\right| \geq |J_l| \geq |J|/k.$$

Hence

$$\left|\bigcup_{j\in J} A_j\right| \geq |J|/k \tag{1}$$

for all $J \subset I$. As we shall see, the trivial necessary condition (1) is also sufficient for the existence of a partition $\mathcal{A} = \bigcup_{i=1}^{k} \mathcal{A}_i$; this was first proved by Edmonds and Fulkerson (1965).

The partition problem above is very close to a problem first considered and solved by Horn (1955). Let U be a finite set of vectors in a vector space V. Under what conditions can U be partitioned as $U = \bigcup_{i=1}^{k} U_i$ such that each set U_i is independent? To state a trivial necessary condition for the existence of such a partition, write $rk(W)$ for the *rank* of W, i.e. for the dimension of the subspace spanned by a set $W \subset V$. Suppose $U = \bigcup_{i=1}^{k} U_i$ with each U_i being independent. Let $W \subset U$ and set $W_i = W \cap U_i$. Then $|W_l| \geq |W|/k$ for some l so $rk(W) \geq rk(W_l) = |W_l| \geq |W|/k$. Hence for all $W \subset U$ we have

$$rk(W) \geq |W|/k. \tag{2}$$

Horn proved that (2) is sufficient for the existence of a partition of U into k independent sets.

Hall's theorem itself has a natural extension to subsets of a vector space. Let $\mathcal{A} = \{A_i : i \in I\}$ be a (finite) family of (finite) subsets of a vector space V. When does $\mathcal{A}$ have an independent transversal? Suppose $T = \{a_i : i \in I\}$ is an independent transversal, i.e. $a_i \in A_i$, $a_i \neq a_j$ if $i \neq j$, and $T \subset V$ is independent. Then for $J \subset I$ we have

$$rk(\bigcup_{j \in J} A_j) \geq rk\{a_j : j \in J\} = |J|.$$

Hence for all $J \subset I$ we have

$$rk(\bigcup_{j \in J} A_j) \geq |J|. \tag{3}$$

Rado (1942) was the first to show that this trivial necessary condition is also sufficient for the existence of an independent transversal.

Our aim in this section is to prove the results above. It so happens that the natural setting for these proofs is *matroid theory*, though we shall hardly need more than the language of matroid theory. The results we shall present are due to Rado (1942), Edmonds (1965), Edmonds and Fulkerson (1965) and Nash-Williams (1967); the interested reader is advised to consult Welsh (1976) for a wealth of additional results.

A *matroid* is a natural generalization of a subset of a vector space: it is a pair $(S, \mathcal{I})$, where S is a set and $\mathcal{I}$ is a set system on S satisfying

(i) if $A \in \mathcal{I}$ and $B \subset A$ then $B \in \mathcal{I}$,

(ii) if $A, B \in \mathcal{I}$ and $|A| > |B|$ then $B \cup \{a\} \in \mathcal{I}$ for some $a \in A \setminus B$.

The sets in $\mathcal{I}$ are said to be *independent sets* and S is the *ground set*.

The matroids we consider are always finite. Note that if S is a set of points in a vector space V and $\mathcal{I}$ is the collection of its independent sets then $(S, \mathcal{I})$ is a matroid. Another kind of matroid can be obtained as follows. Let $\mathcal{A} = \{A_s : s \in S\}$ be a family of not necessarily distinct sets. Call a set $L \subset S$ *independent* if $\{A_l : l \in L\}$ has a transversal. Then S and the collection of independent subsets of S form a matroid, the *transversal matroid* of $\mathcal{A}$ (see Ex. 1). Thus each of the three problems mentioned earlier concerns special matroids.

A matroid $(S, \mathcal{I})$ is determined by its *rank function* $\rho : \mathcal{P} \to \mathbf{Z}^+ = \{0, 1, 2, \ldots\}$, where $\rho(U)$ is the maximal cardinality of an independent subset of U. Clearly,

(i′) $\rho(U) \leq |U|$

(ii′) $\rho(U) \le \rho(W)$ if $U \subset W$,

(iii′) $\rho(U \cup W) + \rho(U \cap W) \le \rho(U) + \rho(W)$.

Conversely, if $\rho : (S) \to \{0, 1, 2, \ldots\}$ is a function satisfying (i)′, (ii)′, (iii)′ then $\mathcal{I} = \{U \subset S : \rho(U) = |U|\}$ is the collection of independent sets of a matroid whose rank function is precisely ρ (Ex. 2). Note also that, corresponding to the exchange axiom (ii), we have that if $\rho(A) > \rho(B)$ then $\rho(B \cup \{a\}) > \rho(B)$ for some $a \in A$.

Let us prove then the matroid version of Hall's theorem, first proved by Rado (1942), and usually called the Rado-Hall theorem.

Theorem 1. *Consider a matroid with ground set S and rank function ρ, and let $\mathcal{A} = \{A_i : i \in I\}$ be a family of (not necessarily distinct) subsets of S. Then $\mathcal{A}$ has an independent transversal iff*

$$\rho\Big(\bigcup_{j \in J} A_j\Big) \ge |J| \tag{4}$$

for all $J \subset I$.

Proof. If $T = \{a_1, a_2, \ldots, a_m\}$, $a_i \in A_i$, is a transversal of $\mathcal{A}$ which is an independent set of the matroid then for every $J \subset I$ we have

$$\rho\Big(\bigcup_{j \in J} A_j\Big) \ge \rho\Big(\bigcup_{j \in J} \{a_j\}\Big) = |J|$$

since $\{a_j : j \in J\}$ is an independent set.

As in all these results, the beauty of the theorem is that the simple necessary condition (4) is, in fact, sufficient for the existence of an independent transversal.

We shall prove by induction on $|I| = m$ that (4) implies the existence of an independent transversal. The assertion being trivial for $m = 1$, assume that $m > 1$ and the result holds for smaller values of m. Let $I = [m]$ and for $U \subset S$ define

$$\rho'(U) = \min\{\rho(U), \rho(U \cup A_1) - 1\}.$$

It is easily seen that ρ' satisfies (i)′, (ii)′ and (iii)′, so it is the rank function of a matroid. Furthermore, by (4), for $J \subset \{2, 3, \ldots, m\}$ we have

$$\rho'\Big(\bigcup_{j \in J} A_j\Big) = \min\Big\{\rho\Big(\bigcup_{j \in J} A_j\Big), \rho\Big(A_1 \cup \bigcup_{j \in J} A_j\Big) - 1\Big\} \ge |J|.$$

By the induction hypothesis the system $\{A_i : 2 \leq i \leq m\}$ has a transversal $T' = \{a_2, a_3, \ldots, a_m\}$ which is independent in the new matroid, i.e. satisfies $\rho'(T') = m - 1$. Hence $\rho(T') \geq |T'|$, implying

$$\rho(T') = |T'| \leq \rho(A_1 \cup T') - 1.$$

Therefore A_1 has an element a_1 such that

$$\rho(T' \cup \{a_1\}) = |T'| + 1.$$

This means that $T = T' \cup \{a_1\}$ is an independent transversal of $\{A_1, A_2, \ldots, A_m\}$. ■

It is easily seen that the proof above gives the following slightly more general result (Ex. 3).

Suppose $\rho : \mathcal{P}(S) \to \mathbf{Z}^+$ satisfies (ii)′ and (iii)′ (such a ρ is called *submodular*) and $\mathcal{A} = (A_1, \ldots, A_m)$ is such that

$$\rho\Big(\bigcup_{j \in J} A_j\Big) \geq |J|$$

for all $J \subset \{1, \ldots, m\}$. Then $\mathcal{A}$ has a system of representatives $(a_1, \ldots, a_m)$ such that $\rho\{a_i : 1 \leq i \leq m\} \geq m$. (Note that the representatives are not demanded to be distinct: the set $\{a_i : 1 \leq i \leq m\}$ may have fewer than m elements, see Ex. 4.)

Formulating Theorem 1 for sets of vectors, we get the following beautiful result mentioned earlier.

Theorem 2. *Let $\{A_i : i \in I\}$ be a finite family of not necessarily distinct finite subsets of a vector space V. Then the family has an independent set of distinct representatives iff*

$$rk\Big(\bigcup_{j \in J} A_j\Big) \geq |J|$$

for all $J \subset I$, where $rk(W)$ is the dimension of the subspace spanned by W. ■

The defect form of Theorem 1 is an easy consequence of the theorem.

Corollary 3. *Let $m \geq 1$, $I = \{1, 2, \ldots, m\}$ and let $\{A_i : i \in I\}$ be a family of (not necessarily distinct) subsets of the ground set S of a matroid with rank function ρ. Then*

$$\max\{k : \textit{some } k \textit{ of the } A_i \textit{ have an independent transversal}\}$$

$$= \min\{|I \setminus J| + \rho(\bigcup_{j \in J}) : J \subset I\}.$$

Proof. If $T = \{a_{i_l} : 1 \le l \le k\}$ is an independent transversal of $\{A_{i_l} : 1 \le l \le k\}$ then

$$\rho(\bigcup_{j \in J} A_j) \ge \left| T \cap (\bigcup_{j \in J} A_j) \right| \ge k + |J| - m$$

for all $J \subset I$.

Conversely, suppose

$$\rho(\bigcup_{j \in J} A_j) \ge |J| - d \tag{5}$$

for all $J \subset I$. We have to show that some $m - d$ of the A_i have an independent transversal. Let $B = \{b_1, \ldots, b_d\}$ be disjoint from S, set $S' = S \cup B$, $A_i' = A_i \cup B$, $1 \le i \le m$, and for $U \subset S'$ define

$$\rho'(U) = \rho(U \setminus B) + |U \cap B|.$$

Then ρ' is the rank function of a matroid on S'. For $J \subset I$ we have, by (5),

$$\rho'(\bigcup_{j \in J} A_j') = (\bigcup_{j \in J} A_j) + d \ge |J|.$$

Hence, by Theorem 1, the family $\{A_i' : i \in I\}$ has a ρ'-independent transversal, say $T' = \{a_1, a_2 \ldots, a_m\}$. Then $T = T' \setminus B$ is a ρ-independent transversal of at least $m - d$ sets A_i.

∎

The proof above becomes slightly easier if instead of Theorem 1 we make use of the statement after Theorem 1, for then we may just take a new submodular function on S, namely $\rho'(Y) = \rho(U) + d$. Of course, this is almost precisely the proof above since we considered only the values

$$\rho'(\bigcup_{j \in J} A_j') = \rho(\bigcup_{j \in J} A_j) + d.$$

Consider a matroid M on a ground set S, with rank function ρ. Call a set *k-independent* if it is the union of k independent sets of M. It is immediate that the k-independent sets form a matroid M_k on S, say with rank function ρ_k, so that $\rho_k(A) = \max\{|\bigcup_{i=1}^k A_i| : \text{each } A_i$ is a ρ-independent subset of $A\}$. The rank formula, due to Edmonds and Fulkerson (1965), is the defect form of a generalization of Horn's theorem (1955); as pointed out by Welsh (1970), it is easily deduced from the defect form of the Rado-Hall theorem.

Theorem 4. *For $A \subset S$ we have*

$$\begin{aligned}\rho_k(A) &= \max\{|\bigcup_{i=1}^k A_i| : \text{ each } A_i \text{ is a } \rho\text{-independent subset of } A\}\\ &= \min\{k\rho(B) + |A \setminus B| : B \subset A\}.\end{aligned}$$

Proof. If $A_1 \ldots, A_k \subset A$ are independent and $B \subset A$ then

$$|\bigcup_{i=1}^k A_i| \le |A \setminus B| + |\bigcup_{i=1}^k (A_i \cap B)| \le |A \setminus B| + k\rho(B).$$

Hence $\rho_k(A)$ is at most as large as claimed.

To prove the essential assertion of the theorem, suppose

$$k\rho(B) + |A \setminus B| \ge t \ge 0 \tag{6}$$

for all $B \subset A$. We have to show that there are ρ-independent subsets of A, say $A_1, \ldots, A_k$, such that $|\bigcup_{i=1}^k A_i| \ge t$.

Take k disjoint copies of M, say $M_1, \ldots, M_k$, on $S_1, \ldots, S_k$, and take the matroid $\tilde{M}$ on $\tilde{S} = \bigcup_{i=1}^k S_i$ in which a set U is independent iff each $U \cap S_i$ is independent in M_i. Write $\tilde{\rho}$ for the rank function of $\tilde{M}$. For $s \in S$ let $s^{(i)}$ be the copy of s in S_i and set $C(s) = \{s^{(1)}, \ldots, s^{(k)}\} \subset \tilde{S}$. Clearly

$$\tilde{\rho}(\bigcup_{s \in B} C(s)) = k\rho(B)$$

for every $B \subset S$. Hence, by (6),

$$\tilde{\rho}(\bigcup_{s \in B} C(s)) + |A \setminus B| \ge t \tag{7}$$

for all $B \subset A$.

Applying Corollary 3 to the sets $C(s)$, $s \in S$, we see that some t of the sets $C(s)$ have an independent transversal in $\tilde{M}$, say $T \subset$

$\tilde{M}$. Let $T \cap S_1 = \{a_1^{(1)}, a_2^{(1)}, \ldots, a_{l_1}^{(1)}\}$, $T \cap S_2 = \{a_{l_1+1}^{(2)}, a_{l_1+2}^{(2)}, \ldots, a_{l_2}^{(2)}\}, \ldots, T \cap S_k = \{a_{l_{k-1}+1}^{(k)}, a_{l_{k-1}+2}^{(k)}, \ldots, a_{l_k}^{(k)}\}$, where $l_k = t$. Set $t_0 = 0$ and $A_i = \{a_{l_{i-1}+1}, a_{l_{i-1}+2}, \ldots, a_{l_i}\}$, $1 \leq i \leq k$. Then each A_i is independent and $\left|\bigcup_{i=1}^k A_i\right| = t$, completing the proof.

■

The covering theorem of Edmonds (1965) is a special case of Theorem 4; as it is very elegant, we state it explicitly. In fact, Theorem 4 is precisely the defect form of Theorem 5.

Theorem 5. *Let ρ be the rank function of a matroid on S. Then S is the union of k independent sets iff*

$$\rho(A) \geq |A|/k$$

for all $A \subset S$. ■

Since the independent sets of a vector space form a matroid, the original form of Horn's theorem (1955) is an immediate consequence of Theorem 5.

Theorem 6. *Let U be a finite set of points in a vector space V. Then U can be partitioned into k independent sets iff*

$$rk(W) \geq |W|/k$$

for all $W \subset V$, where $rk(W)$ is the dimension of the subspace spanned by W. ■

As we have just seen, Horn's theorem is a rather easy consequence of Rado's extension (1942) of Hall's theorem. However, this was not clear at the time Horn discovered his beautiful theorem. Indeed, the result was rediscovered by Rado (1962) who gave essentially the same intricate proof, valid only for vector spaces.

Applying Theorem 5 to transversal matroids we obtain the theorem of Edmonds and Fulkerson (1965).

Theorem 7. *A finite family of finite sets $\{A_i : i \in I\}$ is a union of k families each of which has a transversal iff*

$$\left|\bigcup_{j \in J} A_j\right| \geq |J|/k$$

for all $J \subset I$. ■

Exercises

1. Let $\{A_s : s \in S\}$ be a family of not necessarily distinct sets, where S is a finite index set. Call a set $L \subset S$ *independent* if $\{A_l : l \in L\}$ has a transversal. Show that the independent subsets of S form a matroid. (This matroid is the transversal matroid of the family.)

2. Let S be a finite set and let $\rho : \mathcal{P}(S) \to \mathbf{Z}^+$ satisfy (i)′, (ii)′ and (iii)′. Prove that ρ is the rank function of a matroid on S.
(Set $\mathcal{I} = \{U \subset S : \rho(U) = |U|\}$. Show that $\mathcal{I}$ is the collection of independent sets of a matroid whose rank function is ρ. For $B \subset A \in \mathcal{I}$, by (iii)′ and (i)′, we have $|a| = \rho(A) \leq \rho(B) + \rho(A \setminus B) \leq |B| + |B \setminus A| = |A|$ so $B \in \mathcal{I}$, implying (i). Also, if $\rho(B) = \rho(B \cup \{c\}) = \rho(B \cup \{d\})$ then by (ii)′ and (iii)′, $\rho(B \cup \{c, d\}) = \rho(B)$. This implies (ii) and the fact that ρ is the rank function of the matroid $(S, \mathcal{I})$.)

3. Check that the proof of Theorem 1 gives the slightly more general assertion stated after Theorem 1.

4. Let $S = \{1, 2, 3\}$ and define $\rho : \mathcal{P}(S) \to \mathbf{Z}^+$ by $\rho(A) = 4$ if $1 \in A$ and $\rho(A) = 0$ if $1 \notin A$. Check that ρ satisfies (ii)′ and (iii)′ and $(1, 1, 1, 1)$ is a system of representatives of $\mathcal{A} = \{A \subset S : 1 \in S\}$ satisfying $\rho(\{1\}) = |\mathcal{A}| = 4$. (Note that $\mathcal{A}$ does not have a transversal, i.e. a system of distinct representatives.)

4. Check that the proof of Theorem 4 yields the following theorem of Edmonds and Fulkerson (1965). Let $M_1, \ldots, M_k$ be matroids on the same ground set S, with rank functions $\rho_1, \ldots, \rho_k$. Then

$$\max\Big\{\Big|\bigcup_{i=1}^{k} A_i\Big| : A_i \subset S \text{ is an independent set in } M_i\Big\}$$

$$= \min\Big\{\sum_{i=1}^{k} \rho_i(B_i) + \Big|S \setminus \bigcap_{i=1}^{k} B_i\Big| : B_1, \ldots, B_k \subset S\Big\}.$$

5. A *common transversal* of the set systems $\mathcal{A}$ and $\mathcal{B}$ is, as expected, a set which is both a transversal of $\mathcal{A}$ and a transversal of $\mathcal{B}$. Deduce from Theorem 1 that if $\mathcal{A} = \{A_i : 1 \leq i \leq n\}$ and $\mathcal{B} = \{B_j : 1 \leq j \leq n\}$ then $\mathcal{A}$ and $\mathcal{B}$ have a common transversal iff

$$\Big|\big(\bigcup_{i \in I} A_i\big) \cap \big(\bigcup_{j \in J} B_j\big)\Big| \geq |I| + |J| - n$$

for all $I, J \subset [n]$.

§19. THE FOUR FUNCTIONS THEOREM

In this section we shall touch on a large body of interrelated inequalities. Our main interest in the matter is that many of these inequalities concern pairs of set systems, especially their unions and intersections. The first result in this area is a lemma of Kleitman (1966a) stating that if $\mathcal{A} \subset \mathcal{P}(X)$ is a monotone increasing set system and $\mathcal{B} \subset \mathcal{P}(X)$ is a monotone decreasing set system then

$$|\mathcal{A} \cap \mathcal{B}| \leq 2^{-n}|\mathcal{A}||\mathcal{B}|.$$

This result of Kleitman was the first in a long line of inequalities culminating in the so-called Four Functions Theorem (FFT) of Ahlswede and Daykin (1978). We choose this result as our starting point and deduce a number of its consequences. As is so often the case in combinatorics and, for that matter, in mathematics, the main achievement in obtaining the FFT is finding the statement: having thought of the statement, the proof, though not too short, essentially takes care of itself.

At the first glance the FFT looks too general to be true and, if true, it seems too vague to be of much use. In fact, exactly the opposite is true: the Four Functions Theorem (FFT) of Ahlswede and Daykin is a theorem from "the book". It is beautifully simple and goes to the heart of the matter. Having proved it, we can sit back and enjoy its power enabling us to deduce a wealth of interesting results. This is precisely the reason why this section is rather long: it would be foolish not to present a good selection of the results one can obtain with minimal effort from the FFT. Many of these results are given among the Exercises. The section may be long but very little work is demanded of the reader.

Given set systems $\mathcal{A}, \mathcal{B} \subset \mathcal{P}(X)$, let $\mathcal{A} \vee \mathcal{B} = \{A \cup B : A \in \mathcal{A}, B \in \mathcal{B}\}$ and $\mathcal{A} \wedge \mathcal{B} = \{A \cap B : A \in \mathcal{A}, B \in \mathcal{B}\}$. For a function $\varphi : \mathcal{P}(X) \to \Re$ and a set system $\mathcal{F} \subset \mathcal{P}(X)$, define

$$\varphi(\mathcal{F}) = \sum_{F \in \mathcal{F}} \varphi(F).$$

The Four Functions Theorem of Ahlswede and Daykin (1978) extends an inequality from pairs of subsets of X to pairs of set systems on X. As customary, we write $\Re^+$ for the set of non-negative reals.

Theorem 1. *Let $\alpha, \beta, \gamma, \delta : \mathcal{P}(X) \to \Re^+$ be such that*

$$\alpha(A)\beta(B) \leq \gamma(A \cup B)\delta(A \cap B) \tag{1}$$

for all $A, B \subset X$. Then

$$\alpha(\mathcal{A})\beta(\mathcal{B}) \leq \gamma(\mathcal{A} \vee \mathcal{B})\delta(\mathcal{A} \wedge \mathcal{B}) \tag{2}$$

for all $\mathcal{A}, \mathcal{B} \subset \mathcal{P}(X)$.

Proof. Let us apply induction on n. Let $n = 1$ so that $\mathcal{P}(X) = \mathcal{P}([1]) = \{\emptyset, \{1\}\}$, and for $\epsilon = \alpha, \beta, \gamma$, and δ set $\epsilon_0 = \epsilon(\emptyset)$ and $\epsilon_1 = \epsilon(\{1\})$. Then, by (1),

$$\begin{array}{ll} \alpha_0\beta_0 \leq \gamma_0\delta_0 & \alpha_0\beta_1 \leq \gamma_1\delta_0 \\ \alpha_1\beta_0 \leq \gamma_1\delta_0 & \alpha_1\beta_1 \leq \gamma_1\delta_1 \end{array} \tag{3}$$

If at least one of $\mathcal{A}$ and $\mathcal{B}$ consists of one set only then (2) is easily checked. Suppose then that $\mathcal{A} = \mathcal{B} = \mathcal{P}([1])$. In this case (2) is equivalent to

$$(\alpha_0 + \alpha_1)(\beta_0 + \beta_1) \leq (\gamma_0 + \gamma_1)(\delta_0 + \delta_1) \tag{4}$$

If $\delta_0 = 0$ or $\gamma_1 = 0$ then this is obvious. Otherwise, by (3), we may assume that $\gamma_0 = \alpha_0\beta_0/\delta_0$ and $\delta_1 = \alpha_1\beta_1/\gamma_1$. Then (4) becomes

$$\alpha_1\beta_0 + \alpha_0\beta_1 \leq \gamma_1\delta_0 + \alpha_0\alpha_1\beta_0\beta_1/\gamma_1\delta_0$$

which is equivalent to

$$(\gamma_1\delta_0 - \alpha_0\beta_1)(\gamma_1\delta_0 - \alpha_1\beta_0) \geq 0.$$

This inequality does hold since, by (3), both factors are non-negative.

Let us turn to the induction step. Suppose $n > 1$ and the result holds for smaller values of n. Let us fix two set systems $\mathcal{A}$ and $\mathcal{B}$ on $X = [n]$ and set $Y = [n-1]$. Our aim is to obtain $\alpha(\mathcal{A})$, $\beta(\mathcal{B})$, $\gamma(\mathcal{A} \vee \mathcal{B})$ and $\delta(\mathcal{A} \wedge \mathcal{B})$ from some appropriate functions α', β', γ' and δ' on $\mathcal{P}(Y)$ and then complete the proof by the induction hypothesis.

We may and shall assume that $\alpha = 0$ outside $\mathcal{A}$ and, similarly, $\beta = 0$ outside $\mathcal{B}$, $\gamma = 0$ outside $\mathcal{A} \vee \mathcal{B}$ and $\delta = 0$ outside $\mathcal{A} \wedge \mathcal{B}$. Indeed, if we set $\alpha = 0$ outside $\mathcal{A}$, $\beta = 0$ outside $\mathcal{B}$, etc., then the new functions still satisfy (1), and the values on the appropriate set systems do not change.

What functions α', β', γ' and δ' should we choose on $\mathcal{P}(Y)$? We must choose functions enabling us to recover $\alpha(\mathcal{A})$, $\beta(\mathcal{B})$, $\gamma(\mathcal{A} \vee \mathcal{B})$ and $\delta(\mathcal{A} \wedge \mathcal{B})$. A set $E \subset Y$ is the restriction of two subsets of X to Y, namely $E \subset X$ and $E \cup \{n\} \subset X$. Thus for $\epsilon = \alpha, \beta, \gamma, \delta$ it is natural to try the following functions on $\mathcal{P}(Y)$:

$$\epsilon'(E) = \epsilon(E) + \epsilon(E \cup \{n\}). \tag{5}$$

The choice above implies that with $\mathcal{P}' = \mathcal{P}(Y)$ we have $\alpha(\mathcal{A}) = \alpha(\mathcal{P}) = \alpha'(\mathcal{P}')$, and, similarly, $\beta(\mathcal{B}) = \beta'(\mathcal{P}')$, $\gamma(\mathcal{A} \vee \mathcal{B}) = \gamma'(\mathcal{P}')$ and $\delta(\mathcal{A} \wedge \mathcal{B}) = \delta'(\mathcal{P}')$. Since, trivially, $\mathcal{P}' \vee \mathcal{P}' = \mathcal{P}'$ and $\mathcal{P}' \wedge \mathcal{P}' = \mathcal{P}'$, by the induction hypothesis our proof is complete if we show that the functions α', β', γ' and δ' satisfy (1) on $\mathcal{P}(Y)$. We shall prove this by making use of the case $n = 1$.

Relation (5) strongly suggests that $\epsilon'(E)$ is the value of a function $\overline{\epsilon} : \mathcal{P}([1]) \to \mathfrak{R}^+$ on the full set system $\mathcal{P}([1])$ (admittedly, consisting only of two sets). Let us set then

$$\begin{aligned} \overline{\alpha}(\emptyset) &= \alpha(A) & \overline{\alpha}([1]) &= \alpha(A \cup \{n\}) \\ \overline{\beta}(\emptyset) &= \beta(B) & \overline{\beta}([1]) &= \beta(B \cup \{n\}) \\ \overline{\gamma}(\emptyset) &= \gamma(A \cup B) & \overline{\gamma}([1]) &= \gamma((A \cup B) \cup \{n\}) \\ \overline{\delta}(\emptyset) &= \delta(A \cap B) & \overline{\delta}([1]) &= \delta((A \cap B) \cup \{n\}) \end{aligned}$$

With $\overline{\mathcal{P}} = \mathcal{P}([1])$ we have $\alpha'(A) = \overline{\alpha}(\overline{\mathcal{P}})$, $\beta'(B) = \overline{\beta}(\overline{\mathcal{P}})$, $\gamma'(A \cup B) = \overline{\gamma}(\overline{\mathcal{P}})$ and $\delta'(A \cap B) = \overline{\delta}(\overline{\mathcal{P}})$. Furthermore, inequality (1) for the functions α, β, γ and δ implies that the functions $\overline{\alpha}$, $\overline{\beta}$, $\overline{\gamma}$ and $\overline{\delta}$ on $\mathcal{P}([1])$ also satisfy (1). Hence, as the theorem is true for $n = 1$,

$$\overline{\alpha}(\overline{\mathcal{P}})\overline{\beta}(\overline{\mathcal{P}}) \leq \overline{\gamma}(\overline{\mathcal{P}})\overline{\delta}(\overline{\mathcal{P}})$$

which is precisely the required inequality

$$\alpha'(A)\beta'(B) \leq \gamma'(A \cup B)\delta'(A \cap B). \qquad \blacksquare$$

Recall that a lattice is a partially ordered set in which every pair of elements x, y have a unique least upper bound $x \vee y$ and a unique greatest lower bound $x \wedge y$. A lattice L is *distributive* if for all $x, y, z \in L$ we have

$$x \wedge (y \vee z) = (x \wedge y) \vee (x \wedge z)$$

or, equivalently, for all $x, y, z \in L$,

$$x \vee (y \wedge z) = (x \vee y) \wedge (x \vee z).$$

It is easily proved that every finite distributive lattice is isomorphic to a sublattice of $\mathcal{P}([n])$ for some n (see Birkhoff (1967; p. 59)).

Let us extend the notation $\mathcal{A} \vee \mathcal{B}$, $\mathcal{A} \wedge \mathcal{B}$ to lattices. Given a lattice L and subsets $A, B \subset L$, set

$$A \vee B = \{a \vee b : a \in A, b \in B\}$$

and

$$A \wedge B = \{a \wedge b : a \in A, b \in B\}$$

With this notation Theorem 1 has the following immediate consequence.

Corollary 2. *Let L be a distributive lattice and let $\alpha, \beta, \gamma, \delta : L \to \Re^+$. Then the inequality*

$$\alpha(A)\beta(B) \leq \gamma(A \vee B)\delta(A \wedge B) \tag{6}$$

holds for all subsets $A, B \subset L$ iff it holds for all one-element subsets, i.e. iff

$$\alpha(a)\beta(b) \leq \gamma(a \vee b)\delta(a \wedge b) \tag{7}$$

for all elements $a, b \in L$.

Proof. Consider L embedded in $\mathcal{P}(X)$. Extend each of α, β, γ and δ to the whole of $\mathcal{P}(X)$ by defining them to be 0 outside L. Then (1) is satisfied and the result follows. ∎

We shall list a good selection of closely related and often only slightly different consequences of the results above; the reader should not be too surprised if he can see easy deductions other than those we present.

Inequality (7) does not look too easy to check but in many instances it is entirely trivial. For example, if ϵ is identically 1 on L then $\alpha = \beta = \gamma = \delta = \epsilon$ clearly satisfies (7). Furthermore, $\epsilon(E)$ is simply the number of elements in a set $E \subset L$. The result implied by this choice was first proved by Daykin (1977); it extends Kleitman's lemma we mentioned earlier.

Corollary 3. *If L is a distributive lattice and $A, B \subset L$ then*

$$|A||B| \leq |A \vee B||A \wedge B|.$$ ∎

In fact, as pointed out by Daykin, it is easily seen that a lattice is distributive iff the inequality above holds for all A and B.

In many applications we consider subsets of lattices corresponding to monotone set systems. A subset A of a lattice is said to be a *down-set* if $a \in A$ and $b < a$ imply $b \in A$. Similarly, A is an *up-set* if $a \in A$ and $b > a$ imply $b \in A$. The terminology is not too poetic, but it does convey the meaning without fail. Note that a down-set in the lattice $\mathcal{P}(X)$ is just a monotone decreasing set system on X, and an up-set in $\mathcal{P}(X)$ is a monotone increasing set system on X.

Note that if both A and B are up-sets in a lattice then $A \vee B = A \cup B$, also, if they are both down-sets then $A \wedge B = A \cap B$.

Corollary 4. *Let L be a distributive lattice and let $\mu : L \to \Re^+$ be a function satisfying*

$$\mu(a)\mu(b) \leq \mu(a \vee b)\mu(a \wedge b) \tag{8}$$

for all $a, b \in L$. Then

$$\mu(A)\mu(B) \leq \mu(A \vee B)\mu(A \wedge B) \tag{9}$$

■

Let us reformulate the second assertion in a slightly different form. For a set $E \subset L$ let $\chi_E(x)$ be the characteristic function of E:

$$\chi_E(x) = \begin{cases} 1 & \text{if } x \in E \\ 0 & \text{if } x \in L \setminus E \end{cases}$$

If A and B are both up-sets or both down-sets and μ satisfies (8) then (9) implies that

$$\Big(\sum_{x \in L} \mu(x)\chi_A(x)\Big)\Big(\sum_{x \in L} \mu(x)\chi_B(x)\Big) \leq \Big(\sum_{x \in L} \mu(x)\Big)\Big(\sum_{x \in L} \mu(x)\chi_A(x)\chi_B(x)\Big). \tag{10}$$

From here it is a short step to the celebrated inequality of Fortuin, Kasteleyn and Ginibre (1971), usually referred to as the FKG inequality.

Theorem 5. *Let L be a distributive lattice and suppose $\mu : L \to \Re^+$ satisfies* (8), *i.e.*

$$\mu(x)\mu(y) \leq \mu(x \vee y)\mu(x \wedge y)$$

for all $x, y \in L$. Suppose furthermore that $f, g : L \to \Re^+$ are monotone increasing functions, i.e. if $x < y$ in L then $f(x) \le f(y)$ and $g(x) \le g(y)$. Then we have

$$\Big(\sum_{x\in L} \mu(x)f(x)\Big)\Big(\sum_{x\in L} \mu(x)g(x)\Big) \le \Big(\sum_{x\in L} \mu(x)\Big)\Big(\sum_{x\in L} \mu(x)f(x)g(x)\Big). \quad (11)$$

Proof. All we have to notice is that f and g can be written as positive linear combinations of characteristic functions of up-sets and then apply (10).

Suppose $h : L \to \Re^+$, $h \ne 0$, is a monotone increasing function. Set $S = \operatorname{supp} h = \{x \in L : h(x) > 0\}$, $s = \min\{h(x) : x \in S\} > 0$ and $h' = h - s\chi_S$. Then S is an up-set, h' is monotone increasing and has at least one more zero than h. If $h' \ne 0$ then we can repeat the process and find a non-negative monotone increasing function $h'' = h' - t\chi_T = h - s\chi_S - t\chi_T$ having more zeros than h', where T is an up-set. Continuing in this way, eventually we shall write h as a positive linear combination of characteristic functions of up-sets.

Let then $f = \sum_1^k f_i$ and $g = \sum_1^l g_j$ where the f_i and g_j are positive multiples of characteristic functions of up-sets. By (10) we have

$$\Big(\sum_{x\in L} \mu(x)f_i(x)\Big)\Big(\sum_{x\in L} \mu(x)g_j(x)\Big) \le \Big(\sum_{x\in L} \mu(x)\Big)\Big(\sum_{x\in L} \mu(x)f_i(x)g_j(x)\Big).$$

for all i and j. Inequality (11) is precisely the sum of these inequalities. ∎

The notation suggests that we think of μ as a *measure* on L. Then $\sum_{x\in L} \mu(x)h(x)$ is $\int_L h\,d\mu$. A non-negative function (or measure) μ on a lattice is said to be *log-supermodular* if it satisfies (8). If μ is a non-negative log-supermodular function on a lattice then so is every positive multiple of it. Thus we are allowed to *normalize* μ: we can replace it by a positive multiple of it for which the total mass, $\mu(L)$, is 1. In other words, we may assume that μ is a *probability measure*.

Using this terminology, the FKG inequality states that if μ is a log-supermodular probability measure on a distributive lattice and f and g are non-negative increasing functions then

$$\int f\,d\mu \int g\,d\mu \le \int fg\,d\mu. \quad (12)$$

Let us return to pairs of set systems. Kleitman's lemma, the starting point of all these investigations, can be read out of several results above.

Corollary 6. *Let $\mathcal{A}$, $\mathcal{B}$, $\mathcal{C}$ and $\mathcal{D}$ be set systems on $X = [n]$, with $\mathcal{A}$ and $\mathcal{B}$ monotone decreasing and $\mathcal{C}$ and $\mathcal{D}$ monotone increasing. Then*

$$|\mathcal{A} \cap \mathcal{B}| \geq 2^{-n}|\mathcal{A}||\mathcal{B}|,$$
$$|\mathcal{C} \cap \mathcal{D}| \geq 2^{-n}|\mathcal{C}||\mathcal{D}|$$

and

$$|\mathcal{A} \cap \mathcal{C}| \leq 2^{-n}|\mathcal{A}||\mathcal{C}|$$

Proof. The first two inequalities are immediate from any of the earlier results: from Corollary 3, from inequality (9) with the counting measure ($\mu \equiv 1$), from the FKG inequality, with μ the counting measure and f and g characteristic functions, etc.

To see the third inequality, given $\mathcal{A}$ and $\mathcal{C}$, set $\mathcal{B} = \mathcal{P}(X) \setminus \mathcal{C}$. Then $\mathcal{B}$ is monotone decreasing, $|\mathcal{A} \cap \mathcal{B}| = |\mathcal{A}| - |\mathcal{A} \cap \mathcal{C}|$ and $|\mathcal{B}| = 2^n - |\mathcal{C}|$. Hence, by the first inequality,

$$|\mathcal{A}| - |\mathcal{A} \cap \mathcal{C}| = |\mathcal{A} \cap \mathcal{B}| \geq 2^{-n}|\mathcal{A}|(2^n - |\mathcal{C}|)$$

which gives the third inequality. ∎

The result above has a pleasant reformulation in terms of monotone properties of random sets. Recall from §6 that a property Q of subsets of X is identified with the system of subsets having Q, i.e. $Q \subset \mathcal{P}(X)$. Furthermore, Q is a monotone increasing (decreasing) property if it is monotone increasing (decreasing) as a set system. The probability of Q is simply $P(Q) = 2^{-n}|Q|$. Thus Corollary 6 can be restated as follows.

Corollary 7. *If Q_1 and Q_2 are monotone decreasing properties of X and Q_3 and Q_4 are monotone increasing then*

$$P(Q_1 \cap Q_2) \geq P(Q_1)P(Q_2)$$
$$P(Q_3 \cap Q_4) \geq P(Q_3)P(Q_4)$$
$$P(Q_1 \cap Q_3) \leq P(Q_1)P(Q_3).$$ ∎

This result confirms the intuitively obvious fact that monotone increasing properties are positively correlated while a monotone increasing property and a monotone decreasing property are negatively correlated.

The probability distribution on $\mathcal{P}(X)$ in Corollary 7 is the uniform one: all subsets have the same probability, i.e. $p_A = 2^n$ for all $A \subset X$. Equivalently, to get a random subset, we pick the elements of X independently and with probability $1/2$. In fact, Corollary 7 also holds if a random subset is obtained by selecting the elements of X independently with the same probability p.

The inequalities we have listed so far can often be used to deduce the kind of results we proved about set systems in earlier sections. We know that, trivially, every maximal intersecting family $\mathcal{A} \subset \mathcal{P}(X)$ has 2^{n-1} sets; furthermore, $\mathcal{A}$ is monotone increasing. Also, equally trivially, every maximal family $\mathcal{B} \subset \mathcal{P}(X)$ not containing sets A, B such that $A \cup B = X$ is a monotone decreasing family with 2^{n-1} sets. (The system $\mathcal{P}(X) \setminus \mathcal{B}$ is clearly a maximal intersecting family.) Putting it another way, the probability of a (uniformly distributed random) set belonging to $\mathcal{A}$ is $1/2$ and the probability of belonging to $\mathcal{B}$ is also $1/2$. Therefore, as $\mathcal{A}$ is monotone increasing and $\mathcal{B}$ is monotone decreasing, by Corollary 7 the probability of $\mathcal{A} \cap \mathcal{B}$ is at most $1/4$, i.e. $|\mathcal{A} \cap \mathcal{B}| \leq 2^{n-2}$. This observation, due to Anderson (1976), gives the following result of Seymour (1973), Schönheim (1974) and Daykin and Lovász (1976).

Corollary 8. *Let $\mathcal{F} \subset \mathcal{P}(X)$ be such that if $A, B \in \mathcal{F}$ then $A \cap B \neq \emptyset$ and $A \cup B \neq X$. Then $|\mathcal{F}| \leq 2^{n-2}$.*

Proof. Let $\mathcal{A}$ be a maximal intersecting family containing $\mathcal{F}$ and let $\mathcal{B}$ be a maximal family containing $\mathcal{F}$ without a set in the complement of another. Then, as $\mathcal{F} \subset \mathcal{A} \cap \mathcal{B}$, by the remark above we have $|\mathcal{F}| \leq |\mathcal{A} \cap \mathcal{B}| \leq 2^{n-2}$. ∎

The original purpose of Kleitman (1966a) in proving his lemma (Corollary 6) was to prove the following beautiful result.

Theorem 9. *The union of k intersecting families contains at most $2^n - 2^{n-k}$ sets. The bound is best possible for all k and n, $1 \leq k \leq n$.*

Proof. We apply induction on k. The case $k = 1$ being trivial, we turn to the induction step. Let $\mathcal{F} = \bigcup_1^k \mathcal{F}_i \subset \mathcal{P}(X)$, with each $\mathcal{F}_i$ an intersecting family. Since our aim is to bound $|\mathcal{F}|$ from above, we may assume that each $\mathcal{F}_i$ is a maximal intersecting family. In particular, $|\mathcal{F}_k| = 2^{n-1}$, $\mathcal{A} = \mathcal{P}(X) \setminus \mathcal{F}_k$ is monotone decreasing, $|\mathcal{A}| = 2^{n-1}$, and $\mathcal{B} = \bigcup_1^{k-1} \mathcal{F}_i$ is monotone increasing. By the induction hypothesis $|\mathcal{B}| \leq 2^n - 2^{n-k+1}$ and, by Corollary 6, say,

$$|\mathcal{A} \cap \mathcal{B}| \leq 2^{-n} 2^{n-1}(2^n - 2^{n-k+1}) = 2^{n-1} - 2^{n-k}.$$

therefore

$$|\mathcal{B} \cap \mathcal{F}_k| \geq |\mathcal{B}| - 2^{n-1} + 2^{n-k}.$$

and

$$|\mathcal{F}| = |\bigcup_1^k \mathcal{F}_i| = |\mathcal{B} \cup \mathcal{F}_k| = |\mathcal{B}| + |\mathcal{F}_k| - |\mathcal{B} \cap \mathcal{F}_k| \leq 2^n - 2^{n-k}.$$

The bound is easily seen to be best possible. If $x_1, \ldots, x_k$ denote distinct elements of X let $\mathcal{F}_i = \{A \subset X : x_i \in A\}$. Then each $\mathcal{F}_i$ is intersecting and $\mathcal{F} = \bigcup_1^k \mathcal{F}_i$ contains $2^n - 2^{n-k}$ sets.

With the terminology of §12, the system $\mathcal{F}$ in the proof above is the maximal set system fixed by the sets $\{x_1\}, \ldots, \{x_k\}$. Thus Theorem 9, concerning set systems, is the exact analogue of Theorem 12.3, concerning hypergraphs.

To conclude this section, we deduce a theorem of Marica and Schönheim (1969). Set $\mathcal{A} - \mathcal{B} = \{A \setminus B : A \in \mathcal{A}, B \in \mathcal{B}\}$ for two set systems $\mathcal{A}$ and $\mathcal{B}$.

Theorem 10. *Let $\mathcal{A}, \mathcal{B} \subset \mathcal{P}(X)$. Then*

$$|\mathcal{A} - \mathcal{B}||\mathcal{B} - \mathcal{A}| \geq |\mathcal{A}||\mathcal{B}|.$$

In particular,

$$|\mathcal{A} - \mathcal{A}| \geq |\mathcal{A}|.$$

Proof. For a set system $\mathcal{F}$ let $\overline{\mathcal{F}} = \mathcal{F}^c = \{X \setminus F : F \in \mathcal{F}\}$. Then $|\overline{\mathcal{F}}| = |\mathcal{F}|$. Furthermore,

$$\mathcal{A} - \mathcal{B} = \mathcal{A} \wedge \overline{\mathcal{B}} \qquad \text{and} \qquad \mathcal{B} - \mathcal{A} = \mathcal{B} \wedge \overline{\mathcal{A}}.$$

Consequently, by Corollary 3,

$$\begin{aligned} |\mathcal{A}||\mathcal{B}| = |\mathcal{A}||\overline{\mathcal{B}}| &\leq |\mathcal{A} \vee \overline{\mathcal{B}}||\mathcal{A} \wedge \overline{\mathcal{B}}| = |\overline{\mathcal{A} \vee \overline{\mathcal{B}}}||\mathcal{A} \vee \mathcal{B}| \\ &= |\overline{\mathcal{A}} \wedge \mathcal{B}||\mathcal{A} \vee \overline{\mathcal{B}}| = |\mathcal{B} - \mathcal{A}||\mathcal{A} - \mathcal{B}|. \end{aligned}$$

∎

Daykin and Lovász (1976) and Ahlswede and Daykin (1979) proved generalizations of the second inequality in Theorem 10 (see Exercise 11).

Exercises

1. Check that Corollary 2 does hold for infinite distributive lattices as well.

2. Consider the probability distribution on $\mathcal{P}(X)$ in which the probability of a set $A \subset X$ with k elements is $p_A = p^k(1-p)^{n-k}$. Thus to get a random subset of X, we select the elements independently of each other, with the same probability p. Write $P_p(Q)$ for the probability of an event (property) Q. Deduce from the FKG inequality (Theorem 5) that if Q_1 and Q_2 are monotone decreasing properties of subsets of X and Q_3 and Q_4 are monotone increasing properties then

$$P_p(Q_1 \cap Q_2) \geq P_p(Q_1)P_p(Q_2)$$
$$P_p(Q_3 \cap Q_4) \geq P_p(Q_3)P_p(Q_4)$$
$$P_p(Q_1 \cap Q_3) \leq P_p(Q_1)P_p(Q_3)$$

3. The probability space $\mathcal{G}(n,p) = \mathcal{G}(n, P(\text{edge}) = p)$ of random graphs of order n consists of all graphs on $[n] = \{1, 2, \ldots, n\}$ in which to obtain a random graph we select the edges independently and with the same probability p (see §6). Check that Exercise 2 implies that in $\mathcal{G}(n,p)$ two monotone increasing properties of graphs are positively correlated while a monotone increasing property and a monotone decreasing property are negatively correlated.

4. Let A and B be up-sets in a distributive lattice L and let C be a down-set. Suppose $|B| = |C|$. Prove the following inequality, due to Ahlswede and Daykin (1979):

$$|A \cap B| \geq |A \cap C|.$$

5. Let $(\alpha_i)_1^n$ and $(\beta_i)_1^n$ be sequences of non-negative real numbers. Set $\gamma_k = \max\{\alpha_i\beta_j : \max\{i,j\} = k\}$, $k = 1, 2, \ldots, n$. Prove that

$$\sum_1^n \alpha_i \sum_1^n \beta_j \leq n \sum_1^n \gamma_k.$$

(Consider the distributive lattice $L = [n]$ and for $\epsilon = \alpha, \beta, \gamma$ define $\epsilon : L \to \Re$ by $\epsilon(i) = \epsilon_i$. Furthermore, set $\delta \equiv 1$. Apply Corollary 2.)

6. Deduce from the previous exercise that if $\alpha_1, \ldots, \alpha_n$ and $\beta_1, \ldots, \beta_n$ are increasing sequences of real numbers then

$$n \sum_1^n \alpha_i \beta_{n+1-i} \leq \sum_1^n \alpha_i \sum_1^n \beta_i \leq n \sum_1^n \alpha_i \beta_i.$$

7. Let $\alpha_1, \ldots, \alpha_n$ and $\beta_1, \ldots, \beta_n$ be increasing sequences of real numbers and let $\mu_1, \ldots, \mu_n$ be non-negative reals satisfying $\sum_1^n \mu_i = 1$. Show that

$$\sum_1^n \mu_i \alpha_i \beta_{n+1-i} \leq \Big(\sum_1^n \mu_i \alpha_i\Big)\Big(\sum_1^n \mu_i \beta_i\Big) \leq \sum_1^n \mu_i \alpha_i \beta_i.$$

(Note that every function $\mu : L = [n] \to \Re^+$ satisfies (8) and apply the FKG inequality, Theorem 5)

8. Let μ_1 and μ_2 be probability measures on a finite distributive lattice such that $\mu_1(x)\mu_2(y) \leq \mu_1(x \vee y)\mu_2(x \wedge y)$ for all x, y. Prove Holley's inequality (1974) stating that if f is an increasing function on the lattice then

$$\int f \, d\mu_1 \geq \int f \, d\mu_2.$$

(Assume $f \geq 0$ and set $\alpha(x) = \mu_1(x)$, $\beta(x) = f(x)\mu_2(x)$, $\gamma(x) = f(x)\mu_1(x)$ and $\delta(x) = \mu_2(x)$. Check inequality (7) and apply Corollary 2.)

9. Make use of Theorem 13.4 of Katona (1964) to prove the following result of Anderson (1976), extending Corollary 8. Suppose $n + k$ is odd, say $n + k = 2s + 1$, and $\mathcal{F} \subset \mathcal{P}(X)$ is such that $|A \cap B| \geq k \geq 1$ and $A \cup B \neq X$ for all $A, B \in \mathcal{F}$. Then

$$|\mathcal{F}| \leq \sum_{i=s}^{n-1} \binom{n-1}{i}.$$

(Let $\mathcal{A} = \{A \subset X : A \supset B \text{ for some } B \in \mathcal{F}\}$ and $\mathcal{B} = \{B \subset X : B \subset A$ for some $A \in \mathcal{F}\}$. Then $\mathcal{F} \subset \mathcal{A} \cap \mathcal{B}$, $|\mathcal{B}| \leq 2^{n-1}$ and $|\mathcal{A}|$ is bounded in Theorem 13.4. Apply Kleitman's lemma (Corollary 6) to complete the solution.)

10. Prove the following theorem of Seymour (1973). Suppose $\mathcal{A}, \mathcal{B} \subset \mathcal{P}(X)$ are such that if $A \in \mathcal{A}$ and $B \in \mathcal{B}$ then $A \not\subset B$ and $B \not\subset A$. Then

$$|\mathcal{A}|^{1/2} + |\mathcal{B}|^{1/2} \leq 2^{n/2}.$$

Deduce Corollary 8 from this inequality. (Let $\mathcal{A}_1 = \{A_1 \subset X : A_1 \supset A$ for some $A \in \mathcal{A}\}$, $\mathcal{A}_2 = \mathcal{P}(X)\setminus\mathcal{A}_1$ and define $\mathcal{B}_1$ and $\mathcal{B}_2$ analogously. Then $\mathcal{A} \subset \mathcal{A}_1 \cap \mathcal{B}_2$ and $\mathcal{B} \subset \mathcal{A}_2 \cap \mathcal{B}_1$. Set $a_{ij} = |\mathcal{A}_i \cap \mathcal{B}_j|$ and note that $a_{11} + a_{12} + a_{21} + a_{22} = 2^n$. Check that Corollary 6 implies

$$a_{12}2^n = a_{12}(a_{11} + a_{12} + a_{21} + a_{22}) \leq (a_{11} + a_{12})(a_{12} + a_{22})$$

so

$$a_{12}a_{21} \leq a_{11}a_{22} \leq \left(\frac{a_{11} + a_{22}}{2}\right)^2 = \left(\frac{2^n - a_{12} - a_{21}}{2}\right)^2.$$

Observe that this inequality is precisely the assertion.)

11. Let $\mathcal{A} \subset \mathcal{P}(X)$ be a family of m sets and let $\mathcal{B} \subset \mathcal{P}(X)$ be such that every $A \in \mathcal{A}$ contains some member of $\mathcal{B}$. Prove the theorem of Daykin and Lovász (1976), stating that $\mathcal{A} - \mathcal{B} = \{A\setminus B : A \in \mathcal{A}, B \in \mathcal{B}\}$ contains at least m sets.

(Apply induction on n and proceed as Ahlswede and Daykin (1979). For $F \subset X$ write $F^+ = F \cup \{n\}$ and $F^- = F\setminus\{n\}$. Set $\mathcal{A}_0 = \{A^- : A^-, A^+ \in \mathcal{A}\}$, $\mathcal{A}_1 = \{A^- : A \in \mathcal{A}\}$, $\mathcal{B}_0 = \{B^- : B^- \in \mathcal{B}\}$, $\mathcal{B}_1 = \{B^- : B \in \mathcal{B}\}$ and if $C \in \mathcal{A}_0 - \mathcal{B}_0$ then $C^-, C^+ \in \mathcal{A} - \mathcal{B}$ and if $D \in \mathcal{A}_1 - \mathcal{B}_1$ then at least one of D^- and D^+ belongs to $\mathcal{A} - \mathcal{B}$.)

§20. INFINITE RAMSEY THEORY

This section, the last one in the book, is exceptional in several respects. So far we have considered only finite sets, but here our aim is to study infinite subsets of $\mathbf{N}$. The other sections were practically self-contained, but in order to appreciate this section the reader should be familiar with the rudiments of topology. Furthermore, the methods in this section are rather different from those in the other sections.

Suppose that we are given a 2-colouring of $\mathbf{N}^{(2)}$, i.e. each 2-subset of $\mathbf{N}$ is coloured red or blue, say. Can we find an infinite *monochromatic* subset, i.e. an infinite subset M of $\mathbf{N}$ such that all 2-subsets of M have the same colour?

This is a typical question in Ramsey theory which, loosely speaking, aims at finding some order in a large amount of disorder. In the example above, the colouring of $\mathbf{N}^{(2)}$ is the 'disorder' and the colouring of $M^{(2)}$ is the 'order'. A little more precisely, we start with a very large set carrying some structure, then colour some parts of this structure arbitrarily with a few colours, creating disorder. A Ramsey-type result then claims that no matter how much disorder we tried to create, we can still find a large monochromatic substructure.

In a sense, the very simplest Ramsey-type result is the Dirichlet pigeonhole principle: if the elements of a set of size greater then n are coloured with n colours then there is a monochromatic subset of size 2. This is a frivolous example; here is one that is far from being frivolous. For all natural numbers k and l, there is a natural number $W(k,l)$ such that if $1, 2, \ldots, W(k,l)$ are coloured with k colours then there is a monochromatic arithmetic progression of size l. This is a celebrated theorem of van der Waerden (1927), proved before Ramsey published his fundamental theorems.

In this section we examine various Ramsey theorems concerning colourings of finite subsets of $\mathbf{N}$ and of infinite subsets of $\mathbf{N}$. We shall start off gently, proving Ramsey's (1929) classical theorem about finite subsets of $\mathbf{N}$ and using it to deduce Ramsey's theorem for subsets of finite sets. (Other ways of proving finite Ramsey theorems will be given in the Exercises.) With hindsight, it is clear that Ramsey's theorems opened up wonderful new vistas in mathematics, especially in combinatorics and set theory. Nevertheless, the importance of Ramsey's theorems was recognized only years later, in the 50's and 60's, thanks mostly to the stimulating influence of Erdős. For an excellent account of many results in Ramsey theory, the reader is advised to consult Graham, Rothschild and Spencer (1980).

The major part of the section is a collection of theorems concerning colourings of infinite subsets of $\mathbf{N}$, culminating in a theorem of Ellentuck (1974) describing the completely Ramsey sets. The results we prove are all very pleasingly elegant.

Without further ado, let us return to the first problem above. If we have a very simple colouring of $\mathbf{N}^{(2)}$, for example in which ab is red if $a+b$ is even and blue if $a+b$ is odd, then it is very easy to find an infinite monochromatic subset. However, our task is not always so easy. For example, infinite monochromatic subsets are rather elusive for the colouring in which ab is red if $a+b$ has an even number of prime factors and blue if $a+b$ has an odd number of prime factors. It is a remarkable fact that we can *always* find an infinite monochromatic subset! The proof is short and simple.

Theorem 1. *For every colouring of* $\mathbf{N}^{(2)}$ *with two colours, there is an infinite monochromatic subset of* $\mathbf{N}$.

Proof. Given a colouring of $\mathbf{N}^{(2)}$, choose $a_1 \in \mathbf{N}$ and put $A_1 = \mathbf{N}\setminus\{a_1\}$. Because A_1 is infinite, it contains an infinite subset B_1 such that all pairs a_1b, $b \in B_1$, have the same colour, say colour c_1. Now choose $a_2 \in B_1$ and put $A_2 = B_1\setminus\{a_2\}$. Again, A_2 has an infinite subset B_2 such that all pairs a_2b, $b \in B_2$, have the same colour, c_2, say. Continuing inductively, we obtain a sequence (a_i) in $\mathbf{N}$ and colours (c_i) such that a_ia_j has colour c_i if $i < j$. But the colour sequence (c_i) takes on at most two values so we can find a subsequence $(c_{n(j)})$ which is constant. Then $M = \{a_{n(j)} : j \in \mathbf{N}\}$ is an infinite monochromatic set. ∎

It is easily seen that the same proof applies if we use k colours instead of just 2. (Here, as elsewhere, k is a natural number.) Though the term 'k-colouring' can hardly be misunderstood, let us note that a *k-colouring* of a set S is a map ψ of S into a set $\{c_1, \ldots, c_k\}$ of k *colours.*

Usually we take $1, 2, \ldots, k$ for the colours and if $\psi(x) = l$ then we say that x is *coloured* with l or that the *colour of* x is l. If ψ is a k-colouring of $L^{(r)}$ then $M \subset L$ is said to be *monochromatic* if ψ is constant on $M^{(r)}$.

The k-colour version of Theorem 1 is also an easy consequence of Theorem 1 itself, by induction on k. Given a k-colouring of $\mathbf{N}^{(2)}$ with colours $1, 2, \ldots, k$, Theorem 1 gives us an infinite set $M \subset \mathbf{N}$ such that either $M^{(2)}$ has only colour k or $M^{(2)}$ has only colours $1, 2, \ldots, k-1$. In the first case we are done and in the second we apply the $(k-1)$-colour version of our result to the $(k-1)$-colouring of $\mathbf{M}^{(2)}$ to give us an infinite monochromatic set.

Having seen that k-colouring 2-sets is no more difficult than 2-colouring 2-sets, let us see what happens if we colour the set of r-sets. Of course, for $r = 1$ this is trivial, being the infinite form of the Dirichlet principle, and for $r = 2$ we have just proved it. In fact, a moment's reflection tells us that our proof for $r = 2$ was based on the case $r = 1$. The same idea enables us to deduce the result for r from $r - 1$. This is precisely Ramsey's (1930) fundamental theorem for infinite sets.

Theorem 2. *For every k-colouring of $\mathbf{N}^{(r)}$, there is an infinite monochromatic subset of $\mathbf{N}$.*

Proof. Let us apply induction on r. For $r = 1$ the result is trivial, so we turn to the induction step: let us assume that $r > 1$ and the assertion holds for smaller values of r. Of course, by Theorem 1 and the remarks above we know the result for $r = 2$ as well, but this does not really help us.

Given a k-colouring of $\mathbf{N}^{(r)}$, choose $a_1 \in \mathbf{N}$ and put $A_1 = \mathbf{N} \setminus \{a_1\}$. Induce a k-colouring of $A_1^{(r-1)}$ by colouring $F \in A_1^{(r-1)}$ with the colour of $F \cup \{a_1\}$. By the induction hypothesis, A_1 contains an infinite monochromatic subset B_1 for this colouring. In other words, all sets $F \cup \{a_1\}$, $F \in B_1^{(r-1)}$, have the same colour, say c_1. Now choose $a_2 \in B_1$ and put $A_2 = B_1 \setminus \{a_2\}$. The same argument yields an infinite subset B_2 of A_2 such that all sets $F \cup \{a_2\}$ with $F \in B_2^{(r-1)}$ have the same colour, say c_2.

Continuing in this way, we obtain a sequence of natural numbers $a_1 < a_2 < \ldots$ and a sequence of colours (c_j) such that any set $\{a_{j_1}, \ldots, a_{j_r}\}$ with $j_1 < \ldots < j_r$ has colour c_{j_1}. Taking a subsequence $(c_{n(j)})$ which is constant, we obtain an infinite monochromatic set $\{a_{n(j)} : j \in \mathbf{N}\}$. ∎

It is rather satisfying to deduce the finite version of Ramsey's theorem from the infinite version.

Theorem 3. *Given natural numbers k, r and m, there is a natural number n such that for any k-colouring of $[n]^{(r)}$ there is a monochromatic m-subset of $[n]$.*

Proof. Suppose that, contrary to the assertion, for every $n \in \mathbf{N}$ there is a k-colouring of $[n]^{(r)}$ without a monochromatic m-set. Let $\psi_n : [n]^{(r)} \to [k]$ be such a colouring.

Now, for every n, there are only finitely many ways of k-colouring $[n]^{(r)}$, namely k^N ways, where $N = \binom{n}{r}$. Therefore we can find a subsequence $(\psi_{n_r(j)})$ of our colourings whose terms, when restricted to $[r]^{(r)}$, give the same colouring, say φ_r. Then we can find a subsequence $(\psi_{n_{r+1}(j)})$ of our subsequence $(\psi_{n_r(j)})$ all of whose members restrict to the same colouring φ_{r+1} of $[r+1]^{(r)}$. Next we can find a subsequence $(\psi_{n_{r+2}(j)})$ of $(\psi_{n_{r+1}(j)})$ restricting to the same colouring φ_{r+2} of the set $[r+2]^{(r)}$, and so on. Note that if $s \leq t$ then φ_t restricts to the colouring φ_s of $[s]^{(r)}$.

All that remains is to put together the colourings (ψ_s) to produce a colouring φ of $\mathbf{N}^{(r)}$. What shall we choose for the colour of an r-set $F \in \mathbf{N}^{(r)}$? If $F \subset [s]$ then $\varphi(F) = \varphi_s(F)$ will do. The colour $\varphi(F)$ is well defined since if $F \subset [s] \subset [t]$ then $\varphi_s(F) = \varphi_t(F)$. However, this leads to an emphatic contradiction. By Theorem 2 the colouring φ contains a monochromatic infinite set. On the other hand, φ not only fails to have a monochromatic infinite set but does not even contain a monochromatic m-set. Indeed, if $E \in \mathbf{N}^{(m)}$ and $E \subset [s]$ then φ and φ_s restrict to the same colouring of $[s]^{(r)}$ but E is not monochromatic for φ_s since no m-set is monochromatic for φ_s. ∎

An alternative approach to Theorem 3 goes as follows. A k-colouring of $\mathbf{N}^{(r)}$ is just a point of the product space $K = [k]^{\mathbf{N}^{(r)}}$. Indeed, a point of this product space is precisely a function $\varphi : \mathbf{N}^{(r)} \to [k]$. Give each factor $[k]$ the discrete topology, turning it into a compact space and endow the product K with the product topology. Then by Tychonov's theorem, K is a compact space.

Now, for every finite set $A \subset \mathbf{N}$, the set $C_A = \{\varphi \in K : A$ is not a monochromatic set for $\varphi\}$ is certainly a closed subset of K. If there were no n satisfying the conclusion of Theorem 3 then the family $\{C_A : A \in \mathbf{N}^{(m)}\}$ would have the finite intersection property and so there would be a point $\varphi \in \bigcap\{C_A : A \in \mathbf{N}^{(m)}\}$. This would mean that the colouring φ has no monochromatic m-set — a contradiction, as before.

In fact, these two proofs of Theorem 3 are very much the same: as it happens, in the first proof we just about proved a special case of Tychonov's theorem. Indeed, the reader may mimic the first proof to prove this form of Tychonov's theorem: a countable product of finite spaces is compact (Ex. 1).

Having spent so much time on these proofs of Theorem 3, we must admit that if our main aim had been Theorem 3 itself, we would not have chosen either. The proofs are elegant, no doubt, but give no information whatsoever about the smallest value of n that will do for a given triple r, k and m. Such a proof was given by Erdős and Szekeres (1935) (see Exercises 2 and 3).

So far we have studied only colourings of finite subsets of the same size. Do these results extend to colourings of finite subsets of different sizes? Hardly. If we colour all 1-sets red and all 2-sets blue then there isn't even a monochromatic 2-set. In fact, if we 2-colour the collection of all finite subsets of $\mathbf{N}$ then we cannot even guarantee an infinite set $M \subset \mathbf{N}$ such that for each k all k-sets of M get the same colour, though we can come rather close (see Exercise 5).

Having hit a brick wall here, let us turn to colourings of infinite subsets. Does Theorem 2 extend to colourings of infinite subsets of $\mathbf{N}$? Given a 2-colouring of $\mathbf{N}^{(\omega)}$, the set of all infinite subsets of $\mathbf{N}$, can we always find an $M \in \mathbf{N}^{(\omega)}$ which is monochromatic, i.e. all of whose infinite subsets have the same colour? This is also a tall order and the simple-minded colouring of 1- and 2-sets we have just seen suggests a counterexample. Just make sure that for all $M \in \mathbf{N}^{(\omega)}$ and $x \in M$ the sets M and $M - \{x\}$ get distinct colours. A moment's thought shows that if this condition is satisfied then the colour of M determines the colour of every set L whose symmetric difference with M is finite: L must have the same colour as M if $|M \triangle L|$ is even and the opposite colour if $|M \triangle L|$ is odd.

How can we get a colouring satisfying this condition? Define an equivalence relation $\sim$ on $\mathbf{N}^{(\omega)}$ by putting $L \sim M$ iff $L \triangle M$ is finite. Let $\{E_\gamma : \gamma \in \Gamma\}$ be the set of equivalence classes. Pick a representative M_γ of each equivalence class E_γ. Colour every M_γ red, together with all sets M such that $|M \triangle M_\gamma|$ is even for some M_γ, namely the M_γ in the equivalence class of M, and colour all other infinite subsets blue. Then clearly no $M \in \mathbf{N}^{(\omega)}$ is monochromatic.

Why has this counterexample worked? Because the colour of a set was highly sensitive to the smallest changes in the set. Thus a natural way to proceed is to impose some condition on how fine a colouring we are allowed to take and then see whether we can come up with a monochromatic infinite set. Of course, a colouring is just a splitting of $P = \mathbf{N}^{(\omega)}$ into two subsets, say Y and $Y^c = P \setminus Y$, so what we can hope for is that if we restrict our choice of Y in some sensible way, say in terms of a topology on P, then the colouring (Y, Y^c) will indeed contain an infinite monochromatic set. This is precisely what we shall do — with considerable, perhaps even unexpected, success. The results we shall present are due to Nash-Williams (1965), Mathias (1968), Silver (1970), Galvin and Prikry (1973) and Ellentuck (1974). To be a little more precise: our presentation follows that of Ellentuck (1974) which, in turn, is heavily based on the combinatorial methods of Galvin and Prikry (1973).

Before getting down to some mathematics, let us introduce some terminology and notation. For a set $M \subset \mathbf{N}$, let $M^{(\omega)}$ be the set of all infinite subsets of M and, as before, we set $P = \mathbf{N}^{(\omega)}$. From now on by a *colouring* we understand a 2-colouring of P. Call a colouring *Ramsey* if it contains a monochromatic infinite set. By identifying a set $Y \subset P$ with the colouring (Y, Y^c) it defines, we call Y *Ramsey* if there is a set $M \in \mathbf{N}^{(\omega)}$ such that either $M^{(\omega)} \subset Y$ or $M^{(\omega)} \subset Y^c = P \setminus Y$. The set $\mathcal{P}(\mathbf{N})$ is naturally identified with $2^{\mathbf{N}}$ which, by Tychonov's theorem, is a compact space in the product topology. This gives $P = \mathbf{N}^{(\omega)} \subset \mathcal{P}(\mathbf{N})$ a topology which we shall call the *τ-topology* or the *classical topology* on P. Thus the basic open sets in the classical topology are the sets of the form $\{C \in P : C \cap [n] = A\}$ where $A \subset [n]$.

Let $M^{(<\omega)}$ be the set of all finite subsets of M and set $Q = \mathbf{N}^{(<\omega)}$. For sets $A, B \subset \mathbf{N}$ define $A < B$ if $a < b$ for all $a \in A$ and $b \in B$. Furthermore, set $(A, B)^{(\omega)} = \{C \in \mathbf{N}^{(\omega)} : A \subset C \subset A \cup B \text{ and } A < (C \setminus A)\}$. Thus $(A, B)^{(\omega)}$ is the set of all infinite subsets of $\mathbf{N}$ whose initial segment is A and which continue in B. Note that $P = \mathbf{N}^{(\omega)} = (\emptyset, \mathbf{N})^{(\omega)}$ and $M^{(\omega)} = (\emptyset, M)^{(\omega)}$ for all $M \in P$. Define another topology on P, the *$*$-topology*, by taking as basic open sets the sets of the form $(A, L)^{(\omega)}$ where $A \in Q$ and $L \in P$. If $M \in (A_1, L_1)^{(\omega)} \cap (A_2, L_2)^{(\omega)}$ then $(A_1 \cup A_2, M)^{(\omega)} \subset (A_1, L_1)^{(\omega)} \cap (A_2, L_2)^{(\omega)}$ so the $*$-topology is well-defined. The $*$-topology is finer than the τ-topology because the set $\{(A, \mathbf{N} \setminus [n])^{(\omega)} : A \in Q, n \in \mathbf{N}\}$ is a collection of basic open sets for the τ-topology. The $*$-topology, also called the *Ellentuck topology* or *Mathias topology*, is precisely the topology which is ideal for the study of Ramsey sets.

When studying the Ramsey property, we shall take a set $Y \subset P$, consider it fixed, and then work with various finite and infinite subsets of $\mathbf{N}$, trying to find a suitable subset of P. For sets $M \in P$ and $A \in Q$ we shall say that M *accepts* A (into Y) if $(A, M)^{(\omega)} \subset Y$ and we say that M *rejects* A if there is no $L \in M^{(\omega)}$ such that $(A, L)^{(\omega)} \subset Y$. Note that accepting and rejecting are a bit like being open and closed: for most pairs M, A, the set M neither accepts A nor rejects it. It is clear that if M accepts A then any $L \in M^{(\omega)}$ also accepts A and if M rejects A then any $L \in M^{(\omega)}$ also rejects A. Also, if $A < M$ (as we may always assume) and M accepts A then M accepts any $B \in Q$, $A \subset B \subset A \cup M$.

The key result we need is the following lemma of Galvin and Prikry (1973) which shows that every subset of P comes close to being Ramsey.

Lemma 4. *Let $Y \subset P$ and $M \in P$. If M does not reject $\emptyset$ then some $L \in M^{(\omega)}$ accepts all its finite subsets. If M rejects $\emptyset$ then some $L \in M^{(\omega)}$ rejects all its finite subsets.*

Proof. If M does not reject $\emptyset$ then some $L \in M^{(\omega)}$ accepts $\emptyset$ and so this L accepts all its finite subsets.

Suppose now that M rejects $\emptyset$. We shall construct inductively a sequence $a_1 < a_2 < \dots$ in M rejecting all its finite subsets. Suppose we have chosen $a_1 < a_2 < \dots < a_k$ and $M = M_0 \supset M_1 \supset \dots \supset M_k$ such that $a_i \in M_{i-1}$ and M_i rejects all subsets of $A_i = \{a_1, \dots, a_i\}$. Since $M = M_0$ rejects $\emptyset$, the induction starts.

Suppose then that $k \geq 0$ and M_k and A_k have been constructed but we cannot find M_{k+1} and a_{k+1}. By successively trying various candidates for M_{k+1} and a_{k+1}, we shall show that this assumption leads to a contradiction. Set $N_1 = M_k$ and pick $b_1 \in N_1$ with $b_1 > a_k$. Since $M_{k+1} = N_1$ and $a_{k+1} = b_1$ will not do, there is an $N_2 \in N_1^{(\omega)}$ that accepts some subset F_1 of $A_k \cup \{b_1\}$. As N_2 rejects all subsets of A_k, we must have $F_1 = E_1 \cup \{b_1\}$ for some $E_1 \subset A_k$. Now choose $b_2 \in N_2$, $b_2 > b_1$, and try $M_{k+1} = N_2$ and $a_{k+1} = b_2$. By assumption this will not do either so some $N_3 \in N_2^{(\omega)}$ accepts some subset F_2 of $A_k \cup \{b_2\}$ and so $F_2 = E_2 \cup \{b_2\}$ for some $E_2 \subset A_k$. Then we choose $b_3 \in N_3$, $b_3 > b_2$, and try $M_{k+1} = N_3$ and $a_{k+1} = b_3$. As before, we conclude that some $N_4 \in N_3^{(\omega)}$ accepts some subset F_3 of $A_k \cup \{b_3\}$ and $F_3 = E_3 \cup \{b_3\}$ for some $E_3 \subset A_k$.

Continuing in this way, we find sequences $a_k < b_1 < b_2 < \dots$, $M_k = N_1 \supset N_2 \supset \dots$ and a sequence (E_i) of subsets of A_k such that $b_i \in N_i$, $i = 1, 2, \dots$, and N_{i+1} accepts $E_i \cup \{b_i\}$. There are only finitely many choices for the sets E_i so, by passing to a subsequence, we may assume that all the E_i are the same, say $E_i = E \subset A_k$ for all i. But then $B = \{b_1, b_2, \dots\} \subset M_k$ accepts $E \subset A_k$, contradicting our assumption.

Therefore there are infinite sequences $a_1 < a_2 < \ldots$ and $M = M_0 \supset M_1 \supset \ldots$ such that $a_i \in M_{i-1}$ and M_i rejects all subsets of $A_i = \{a_1, \ldots, a_i\}$. Hence $L = \{a_1, a_2, \ldots\}$ rejects all its finite subsets.

■

From Lemma 4 it is easy to prove that the $*$-open sets are Ramsey, and so, in particular, the classical open sets are Ramsey, as first proved by Nash-Williams (1965).

Theorem 5. *Every $*$-open subset of $P = \mathbf{N}^{(\omega)}$ is Ramsey.*

Proof. Let $Y \subset P$ be $*$-open. Apply Lemma 4 to Y and $M = \mathbf{N}$ to obtain a set $L \in P$. Suppose $L^{(\omega)} \not\subset Y^c$, say $N \in L^{(\omega)} \cap Y$. Since Y is $*$-open, there is an $A \in Q$ such that $N \in (A, N)^{(\omega)} \subset L^{(\omega)} \cap Y \subset Y$. Then N accepts A so L does not reject $A \in L^{(<\omega)}$. Therefore L accepts all its finite subsets, i.e. $L^{(\omega)} = (\emptyset, L)^{(\omega)} \subset Y$. ■

At this point we strengthen the Ramsey property to bring it closer in spirit to the $*$-topology. We say that a set $Y \subset P = \mathbf{N}^{(\omega)}$ is *completely Ramsey* if for all $M \in P$ and $A \in Q$ there is an $L \in M^{(\omega)}$ with $(A, L)^{(\omega)} \subset Y$ or $(A, L)^{(\omega)} \subset Y^c$. Putting $M = \mathbf{N}$ and $A = \emptyset$ in this definition, we see that completely Ramsey sets are Ramsey. Not surprisingly, not every Ramsey set is completely Ramsey. For example, if Y is the non-Ramsey set given earlier then it is easy to see that $Y \cup \{M \in P : 1 \notin M\}$ is Ramsey but not completely Ramsey: if $A = [1]$ then neither $(A, L)^{(\omega)} \subset Y$ nor $(A, L)^{(\omega)} \subset Y^c$ holds for any $L \in P$.

Thanks to the fact that the completely Ramsey property and the $*$-topology are so similar in character, we shall eventually be able to characterize the completely Ramsey subsets of $\mathbf{N}^{(\omega)}$ in terms of the $*$-topology. First we strengthen Theorem 5.

Theorem 6. *Every $*$-open subset of $P = \mathbf{N}^{(\omega)}$ is completely Ramsey.*

Proof. Let $A \in Q$ and $M \in P$. We have to look for a set $K \in M^{(\omega)}$ such that

$$(A, K)^{(\omega)} \subset Y \qquad \text{or} \qquad (A, K)^{(\omega)} \subset Y^c.$$

We shall achieve this by constructing a continuous map $h : P \to (A, M)^{(\omega)}$ mapping every set of the form $L^{(\omega)}$ into a set of the form $(A, L')^{(\omega)}$, where $L' \in M^{(\omega)}$. Having constructed such a map h, it is a simple matter to complete the proof by applying Theorem 5 to the $*$-open set $h^{-1}(Y)$. For if $L^{(\omega)} \subset h^{-1}(Y)$ then $h(L^{(\omega)}) = (A, L')^{(\omega)} \subset Y$ and if $L^{(\omega)} \subset P \backslash h^{-1}(Y)$ then $h(L^{(\omega)}) = (A, L')^{(\omega)} \subset Y^c$.

Let us see then how we can carry out this simple program. Let $\tilde{f}_M$ be the strictly increasing function mapping $\mathbf{N}$ onto M, let $f_M : P \to P$ be the induced map, and let $g_A : P \to P$ be defined by $g_A(L) = A \cup L$, $L \in P$. We claim that f_M and g_A are continuous. Consider any neighbourhood $U = (B, K)^{(\omega)}$, where we may assume that $B < K$. If $f_M(L) \in U$ then $\left(\tilde{f}_M^{-1}(B), \tilde{f}_M^{-1}(K)\right)^{(\omega)}$ is a neighbourhood of L whose image under f_M is contained in U.

Furthermore, if $g_A(L) = A \cup L \in U$ then $(B \cap L, K)^{(\omega)}$ is a neighbourhood of L whose image under g_A is contained in U. This is true because

$$(B \cap L) \cup A = (B \cup A) \cap (L \cup A) \supset B.$$

Thus f_M and g_A are continuous, as claimed.

The map $h = g_A f_M : P \to (A, M)^{(\omega)}$ is continuous and if $L \in P$ then $h(L^{(\omega)}) = (A, f_M(L))^{(\omega)}$, so our proof is complete. ■

It is worth remarking that Theorem 6 is equivalent to saying that the $*$-closed subsets of $\mathbf{N}^{(\omega)}$ are completely Ramsey. Indeed, a set is (completely) Ramsey iff its complement is (completely) Ramsey.

Having looked at the open sets in the $*$-topology, which, in some sense, are the 'large' sets, let us turn our attention to the 'small' sets in the $*$-topology, namely to the nowhere dense sets. Recall that a set is *nowhere dense* if it is not dense in any non-empty open set. Equivalently, a set Y is nowhere dense if every non-empty open set contains a non-empty open set disjoint from Y. Thus Y is nowhere dense iff its closure $\overline{Y}$ is also nowhere dense iff the interior of $\overline{Y}$ is empty. Interpreting this in the $*$-topology, we see that $Y \subset P = \mathbf{N}^{(\omega)}$ is nowhere dense iff for every basic $*$-open set $(A, M)^{(\omega)}$ there is a neighbourhood $(B, L)^{(\omega)} \subset (A, M)^{(\omega)}$ with $(B, L)^{(\omega)} \subset Y^c = P \backslash Y$.

It turns out that all $*$-nowhere-dense sets are completely Ramsey. Even more, they are precisely those completely Ramsey sets for which the second alternative holds in the definition of a completely Ramsey set Y, namely $(A, L)^{(\omega)} \subset Y^c$.

Theorem 7. *A set $Y \subset P$ is nowhere dense in the $*$-topology iff for all $M \in P$ and $A \in Q$ there is an $L \in M^{(\omega)}$ with $(A, L)^{(\omega)} \subset Y^c$.*

Proof. Let $Y \subset P$ be nowhere dense in the $*$-topology. Then $\overline{Y}$ is also nowhere dense and, by Theorem 6, it is completely Ramsey. Hence for $M \in P$ and $A \in Q$ there is an $L \in M^{(\omega)}$ such that either $(A,L)^{(\omega)} \subset \overline{Y}$ or $(A,L)^{(\omega)} \subset P \backslash \overline{Y}$. Now the first alternative cannot hold because $(A,L)^{(\omega)}$ is a non-empty open set but the interior of $\overline{Y}$ is empty. Hence $(A,L)^{(\omega)} \subset P \backslash \overline{Y} \subset P \backslash Y = Y^c$.

The converse implication is trivial: if every basic $*$-open set contains a non-empty $*$-open set disjoint from Y then Y is $*$-nowhere-dense. ∎

After our success with the small sets, let us look at some potentially larger sets, namely the $*$-meagre sets. A set is *meagre* if it is a countable union of nowhere dense sets.

Here the $*$-topology springs a surprise: a set which is $*$-meagre is, in fact, $*$-nowhere-dense.

Theorem 8. *If $Y \subset P$ is $*$-meagre then for all $M \in P$ and $A \in Q$ there is an $L \in M^{(\omega)}$ with $(A,L)^{(\omega)} \subset Y^c$. In particular, every $*$-meagre set is $*$-nowhere-dense.*

Proof. Let $Y = \bigcup_{i=1}^{\infty} Y_i$, with each Y_i $*$-nowhere dense. Given $M \in P$ and $A \in Q$, let us find inductively elements $x_1 < x_2 < \ldots$ in M such that $L = \{x_1, x_2, \ldots\}$ will do: $(A,L)^{(\omega)} \subset Y^c$. As Y_1 is $*$-nowhere-dense, by Theorem 7 there is an $M_1 \in M^{(\omega)}$ such that $(A, M_1)^{(\omega)} \subset Y^c$. Pick $x_1 \in M$ such that $x_1 > \max A$.

Suppose inductively that we have chosen $\max A < x_1 < \ldots < x_k$ and $M \supset M_1 \supset \ldots \supset M_k$ with $x_i \in M_i$ and $(A \cup F, M_i)^{(\omega)} \subset Y_i^c$ for every subset F of $\{x_1, \ldots, x_{i-1}\}$. If we apply Theorem 7 once for each subset F of $\{x_1, \ldots, x_k\}$, so altogether 2^k times, to Y_{k+1} and the sets $A \cup F$, we obtain $M_{k+1} \in M_k^{(\omega)}$ such that $(A \cup F, M_{k+1})^{(\omega)} \subset Y_{k+1}^c$ for all $F \subset \{x_1, \ldots, x_k\}$. Now just pick $x_{k+1} \in M_{k+1}$ with $x_{k+1} > x_k$.

We thus obtain a sequence $x_1 < x_2 < \ldots$ such that, for every i, we have $(A \cup F, \{x_i, x_{i+1}, \ldots\})^{(\omega)} \subset Y_i^c$ for every $F \subset \{x_1, \ldots, x_{i-1}\}$. Then the set $L = \{x_1, x_2, \ldots, \}$ clearly satisfies $(A,L)^{(\omega)} \subset Y^c$. ∎

We can now prove the theorem of Ellentuck (1974) which gives a precise characterization of the completely Ramsey subsets of $P = \mathbf{N}^{(\omega)}$. We need just one more definition from topology: a set has the *Baire property* if it is the symmetric difference of an open set and a meagre set.

Theorem 9. *A subset of P is completely Ramsey iff it has the Baire property in the $*$-topology.*

Proof. (i) Let $Y \subset P$ have the $*$-Baire property: say $Y = Y_1 \triangle Y_2$, where Y_1 is $*$-open and Y_2 is $*$-meagre. Let us check that Y is completely Ramsey. Given $M \in P$ and $A \in Q$, by Theorem 6 we can choose $L_1 \in M^{(\omega)}$ with $(A, L_1)^{(\omega)} \subset Y_1$ or $(A, L_1)^{(\omega)} \subset Y_1^c$. Furthermore, by Theorem 8 we can choose $L_2 \in L_1^{(\omega)}$ such that $(A, L_2)(\omega) \subset Y_2^c$. If $(A, L_1)^{(\omega)} \subset Y_1$ then $(A, L_2)^{(\omega)} \subset (A, L_1)^{(\omega)} \subset Y_1$ and so $(A, L_2)^{(\omega)} \subset Y_1 \cap Y_2^c \subset Y$. Similarly, if $(A, L_1)^{(\omega)} \subset Y_1^c$ then $(A, L_2)^{(\omega)} \subset (A, L_1)^{(\omega)} \cap Y_2^c \subset Y_1^c \cap Y_2^c \subset Y^c$. Thus Y is completely Ramsey.

(ii) Let now $Y \subset P$ be completely Ramsey. We claim that $Y' = Y \setminus (\text{Int } Y)$ is $*$-nowhere dense and so $Y = \text{Int } Y \triangle Y'$ has the $*$-Baire property. To prove our claim, take a basic open set $U = (A, M)^{(\omega)}$. As Y is completely Ramsey, there is an open set $V = (A, L)^{(\omega)} \subset U$ such that $V \subset Y$ or $V \subset Y^c$. Now if $V \subset Y$ then $V \subset \text{Int } Y$ so $V \cap Y' = \emptyset$ and if $V \subset Y^c$ then $V \cap Y' \subset V \cap Y = \emptyset$. Hence $V \subset U$ is an open set disjoint from Y', proving our claim and completing the proof. ∎

We remark that without Theorem 8, we could have concluded that a set is completely Ramsey iff it is the symmetric difference of a $*$-open set and a $*$-nowhere-dense set. However, the importance of knowing that the completely Ramsey sets are those with the $*$-Baire property is that, as is easily seen, the Baire sets form a σ-algebra. This allows us to conclude that the $*$-Borel sets (i.e. sets in the σ-algebra generated by the open sets) are completely Ramsey, so, *a fortiori*, the classical Borel sets are Ramsey. This is precisely the result which is often used in applications of Ramsey theory to analysis.

Corollary 10. *The $*$-Borel sets and the classical Borel sets of $\mathbf{N}^{(\omega)}$ are completely Ramsey.* ∎

Exercises

1. Let (n_i) be a sequence of natural numbers. Give $[n_i]$ the discrete topology and take $P = \prod_1^\infty [n_i]$ with the product topology. Imitate the proof of Theorem 3 to show that P is compact.

2. Let $R_k^{(r)}(s_1, \ldots, s_k)$ be the minimal integer R for which every k-colouring of $R^{(r)}$ with colours $c_1, \ldots, c_k$ contains a monochromatic s_i-subset of colour c_i for some i. Note that the function $R_k^{(r)}(s_1, \ldots, s_k)$ is independent of the order of the s_i and if $s_1 < r \leq s_2 \leq \ldots \leq s_r$ then

$$R_k^{(r)}(s_1, \ldots, s_k) = s_1$$

$$R_k^{(r)}(r, s_2, \ldots, s_k) = R_{k-1}^{(r)}(s_2, \ldots, s_k)$$
$$R_2^{(r)}(r, s_2) = R_1^{(r)}(s_2) = s_2$$

and

$$R_1^{(1)}(s_1, s_2) = s_1 + s_2 - 1.$$

By inducing a k-colouring of the $(r-1)$-subsets of a set $A = B \setminus \{a\}$ from a k-colouring of the r-subsets of B, prove the following inequality, due to Erdős and Szekeres (1935): if $s_i > r$ for all i then

$$R_k^{(r)}(s_1, \ldots, s_k) \le R_k^{(r-1)}\{R^{(r)}(s_1 - 1, s_2, \ldots, s_k),$$

$$R_k^{(r-1)}(s_1, s_2 - 1, s_3, \ldots, s_k), \ldots, R_k^{(r-1)}(s_1, s_2, \ldots, s_{k-1}, s_k - 1)\} + 1.$$

3. Deduce from the result in Exercise 2 that if $s_1, s_2 > 2$ then

$$R(s_1, s_2) \le R(s_1 - 1, s_2) + R(s_1, s_2 - 1)$$

where, as usual, $R_2^{(2)}$ has been abbreviated to R. Hence show that

$$R(s_1, s_2) \le \binom{s_1 + s_2 - 2}{s_1 - 1}.$$

4. Show that $R(3,3) = 6$ and, by considering a star-shaped graph on 8 vertices in which every vertex has degree 3, prove that $R(3,4) = 9$.

5. Let $k_1, k_2, \ldots$ and $l_1, l_2, \ldots$ be sequences of integers such that $l_r \to \infty$ (say $k_r = 2^r$ and $l_r = \lfloor \log\log(r+1) \rfloor$). Show that, for any colouring of $\mathbf{N}^{(<\omega)}$ which uses at most k_r colours for $\mathbf{N}^{(r)}$, there is an infinite set $M = \{m_1, m_2, \ldots\}$, $m_1 < m_2 < \ldots$, such that if $F_1, F_2 \in M^{(r)}$ and $\min(F_1 \cup F_2) \ge m_{l_r}$ then F_1 and F_2 have the same colour.

Show also that the condition $l_r \to \infty$ cannot be omitted.

6. Let $\{a(i_1, \ldots, i_k) : 1 \le i_1 < i_2 < \ldots < i_k, i_j \in \mathbf{N}, k = 1, 2, \ldots\}$ be an array of real numbers such that

$$|a(i_1, \ldots, i_k)| \le B_k$$

for all $i_1 < \ldots < i_k$, $k = 1, 2, \ldots$. Show that there are constants $a_1, a_2, \ldots$ and a set $M = \{m_1, m_2, \ldots\} \in \mathbf{N}^{(\omega)}$ such that

$\lim_{i_1 \to \infty} a(m_{i_1}, m_{i_2}, \ldots, m_{i_k}) = a_k$ for all $k \in \mathbf{N}$, i.e. if $\epsilon > 0$ and $k \in \mathbf{N}$ then

$$\left| a(m_{i_1}, m_{i_2}, \ldots, m_{i_k}) - a_k \right| < \epsilon$$

provided $m_{i_1} < \ldots < m_{i_k}$ and i_1 is sufficiently large.

7. Let $f_1, f_2, \ldots$ be bounded real functions on some set W. Show that there are functions $h_n : \Re^n \to \Re$, $n = 1, 2, \ldots$, and a set $M = \{m_1, m_2 \ldots\} \in \mathbf{N}^{(\omega)}$, such that if $x_1, \ldots, x_n \in \Re$ then

$$\lim_{i_1 \to \infty} \sup_{w \in W} \left| \sum_{j=1}^{n} x_{i_j} f_{m_{i_j}}(w) \right| = h_n(x_1 \ldots, x_n).$$

8. Let $f_1, f_2, \ldots$ be bounded complex functions on some set W. Show that there is a set $L \in \mathbf{N}^{(\omega)}$ such that one of the following two statements holds:

(a) if $M = \{m_1, m_2, \ldots\} \in L^{(\omega)}$ then $\sup_w \left| \sum_1^n f_{m_i}(w) \right| \geq 1/n$ for every $n \geq 1$,

(b) if $M = \{m_1, m_2, \ldots\} \in L^{(\omega)}$ then $\sup_w \left| \sum_1^n f_{m_i}(w) \right| < 1/n$ for some $n \geq 1$.

9. Let $f_1, f_2, \ldots$ be complex functions on some set W and let D_1 and D_2 be disjoint sets of complex numbers. Suppose that for every $L \in P = \mathbf{N}^{(\omega)}$ there is a point $w \in W$ such that $f_n(w) \in D_1$ for infinitely many $n \in L$. Let $Y_k = \{M = \{m_1, m_2, \ldots\} \in P :$ for some $w \in W$ we have $f_{m_i}(w) \in D_1$ if $i \leq k$ and i is odd, and $f_{m_i}(w) \in D_2$ if $i \leq k$ and i is even$\}$, and set $Y = \bigcap_1^\infty Y_k$. Check that Y is closed in the product topology on P.

Deduce that there is a set $M = \{m_1, m_2 \ldots\} \in P$ such that if $\epsilon_i = 1$ or -1, $i = 1, 2, \ldots$, and $k \in \mathbf{N}$ then there is a $w_k \in W$ such that

$$f_{m_i}(w_k) \in D_1 \qquad \text{if } i \leq k \text{ and } \epsilon_i = 1$$

and

$$f_{m_i}(w_k) \in D_2 \qquad \text{if } i \leq k \text{ and } \epsilon_i = -1.$$

REFERENCES

Ahlswede, R. and Daykin, D. E. (1978), An inequality for the weights of two families of sets, their unions and intersections, *Z. Wahrscheinl. Geb.* **43**, 183–185.

Ahlswede, R. and Daykin, D. E. (1979), The number of values of combinatorial functions, *Bull. London Math. Soc.* **11**, 49–51.

Alon, N. (1983), On the density of sets of vectors, *Discrete Math.* **46**, 199–202.

Alon, N. (1986), An extremal problem for sets with applications to graph theory, to appear

Alon, N. and Milman, V. D. (1983), Embedding of l_∞^k in finite dimensional Banach spaces, *Israel J. Math.* **45**, 265–280.

Anderson, I. (1976), Intersection theorems and a lemma of Kleitman, *Discrete Math.* **16**, 181–185.

Asimow, L. and Roth, B. (1978), The rigidity of graphs, *Trans. Amer. Math. Soc.* **245**, 279–289.

Baranyai, Z. (1975), On the factorisation of the complete uniform hypergraph, in *Infinite and finite sets*, (A. Hajnal, T. Rado and V. T. Sós, eds), North-Holland, Amsterdam, pp. 91–108.

Baranyai, Z. (1979), The edge-colouring of complete hypergraphs I, *J. Combinatorial Theory (B)* **26**, 276–294.

Baumert, L. D., McEliece, R. J., Rodemich, E. R. and Rumsey, H. (1980), A probabilistic version of Sperner's theorem, *Ars Combin.* **9**, 91–100.

Bernstein, A. J. (1967), Maximally connected arrays on the n-cube, *SIAM J. Appl. Math.* **15**, 1485–1489.

Birkhoff, G. (1967), *Lattice Theory*, 3rd Edn., American Math. Soc. Coll. Publ. vol. XXV, Amer. Math. Soc., Providence, R.I., $vi + 418$pp.

Bollobás, B. (1965), On generalized graphs, *Acta Math. Acad. Sci. Hungar.* **16**, 447–452.

Bollobás, B. (1973), Sperner systems consisting of pairs of complementary subsets, *J. Combinatorial Theory (A)* **15**, 363–366.

Bollobás, B. (1974), Three-graphs without two triples whose symmetric difference is contained in a third, *Discrete Math.* **8**, 21–24.

Bollobás, B. (1978), *Extremal Graph Theory*, LMS Monographs, No. 11, Academic Press, London, New York, San Francisco, $xx + 488$pp.

Bollobás, B. (1979), *Graph Theory -An Introductory Course*, Graduate Texts in Mathematics, Springer-Verlag, New York, Heidelberg, Berlin, $x + 177$pp.

Bollobás, B. (1985), *Random Graphs*, Academic Press, London, $xvi +$ 447pp.

Bollobás, B. and Duchet, P. (1979), Helly families of maximal size, *J. Combinatorial Theory (A)* **26**, 197–200.

Bollobás, B. and Duchet, P. (1983), On Helly families of maximal size, *J. Combinatorial Theory (B)* **35**, 290–296.

Bollobás, B. and Thomason, A. G. (1985), Threshold functions, *Combinatorica*, to appear

Bondy, J. A. (1972), Induced subsets, *J. Combinatorial Theory (B)*, **12**, 201–202.

Brace, A. and Daykin, D. E. (1972), Sperner type theorems for finite sets, *Proc. British Comb. Conf. Oxford*, 1972, 18–37.

de Bruin, N. G., Tengbergen, C. and Kruyswijk, D. (1952), On the set of divisors of a number, *Nieuw. Arch. Wisk. (2)* **23**, 191–193.

de Caen, D. (1983), A note on the probabilistic approach to Turán's problem, *J. Combinatorial Theory (B)* **34**, 340–349.

Całczyńska-Karłowicz (1964), Theorem on families of finite sets, *Bull. Acad. Polon. Sci., Ser. Math. Astr. Phys.* **12**, 87–89.

Cayley, A. (1850), On the triadic arrangements of seven and fifteen things, *London, Edinburgh and Dublin Phil. Mag. and J. Science* **37**, 50–53.

Clements, G. F. (1973), A minimization problem concerning subsets of a finite set, *Discrete Math.* **4**, 123–128.

Clements, G. F. and Lindström (1969), A generalization of a combinatorial theorem of Macaulay, *J. Combinatorial Theory* **7**, 230–238.

Corrádi, K. A. and Kátai, I. (1969), A note on combinatorics, *Annales Universitatis Scientarum Budapestinensis de Rolando Eötvös Nominatae* **12**, 100–106.

Crawley, P. and Dilworth, R. P. (1973), *Algebraic Theory of Lattices*, Prentice-Hall, Englewood Cliffs.

Daykin, D. E. (1974), A simple proof of the Kruskal-Katona theorem, *J. Combinatorial Theory (A)* **17**, 252–253.

Daykin, D. E. (1977), A lattice is distributive iff $|A||B| \leq |A \vee B||A \wedge B|$, *Nanta Math.* **10**, 58–60.

Daykin, D. E., Godfrey, J. and Hilton, A. J. W. (1974), Existence theorems for Sperner families, *J. Combinatorial Theory (A)* **17**, 245–251.

Daykin, D. E. and Lovász, L. (1976), The number of values of a Boolean function, *J. London Math. Soc.* (2) **12**, 225–230.

Deza, M. and Frankl, P. (1983), Erdős-Ko-Rado theorem — 22 years later, *SIAM J. Alg. Disc. Meth.* **4**, 419–431.

Dilworth, R. P. (1950), A decomposition theorem for partially ordered sets, *Ann. Math.* **51**, 161–165.

Edmonds, J. (1965), Lehman's switching game and a theorem of Tutte and Nash-Williams, *J. Res. Nat. Bur. Stand.* **69B**, 73–77.

Edmonds, J. and Fulkerson, D. R. (1965), Transversals and matroid partition, *J. Res. Nat. Bur. Stand.* **69B**, 147–153.

Ehrenfeucht, A. and Mycielski, J. (1974), On families of intersecting sets, *J. Combinatorial Theory (A)* **17**, 259–260.

Ellentuck, E. E. (1974), A new proof that analytic sets are Ramsey, *J. Symbolic Logic* **39**, 163–165.

Erdős, P. (1945), On a lemma of Littlewood and Offord, *Bull. Amer. Math. Soc.* **51**, 898–902.

Erdős, P. (1965), A problem on independent r-tuples, *Ann. Univ. Sci. Budapest* **8**, 93–95.

Erdős, P., Chao Ko and Rado, R. (1961), Intersection theorems for systems of finite sets, *Quart. J. Math. Oxford (2)* **12**, 313–320.

Erdős, P. and Kleitman, D. J. (1974), Extremal problems among subsets of a set, *Discrete Math.* **8**, 281–294.

Erdős, P. and Szekeres, G. (1935), A combinatorial problem in geometry, *Compositio Math.* **2**, 463–470.

Fortuin, C. M., Kasteleyn, P. W. and Ginibre, J. (1971), Correlation inequalities on some partially ordered sets, *Comm. Math. Phys.* **22** 89–103.

Frankl, P. (1978), The Erdős-Ko-Rado theorem is true for $n = ckt$, *Proc. Fifth Hungarian Comb. Coll.*, North-Holland, Amsterdam, 365–375.

Frankl, P. (1982), An extremal problem for two families of sets, *European J. Combin.* **3**, 125–127.

Frankl, P. (1983), On the trace of finite sets, *J. Combinatorial Theory (A)* **34**, 41–45.

Frankl, P. (1984), A new short proof of the Kruskal-Katona theorem, *Discrete Math.* **48**, 327–329.

Frankl, P. and Füredi, Z. (1981), A short proof for a theorem of Harper about Hamming spheres, *Discrete Math.* **34**, 311–313.

Frankl, P. and Rödl, V. (1986), Forbidden intersections (to appear)

Galvin, F. and Prikry, K, (1973), Borel sets and Ramsey's theorem, *J. Symbolic Logic* **38**, 193–198.

Graham, R. L. (1982), Linear extensions of partial orders and the FKG inequality, in *Ordered Sets* (I. Rival, ed.) D. Reidel, Dordrecht, Holland, 213–236.

Graham, R. L., Rothschild, B. L. and Spencer, J. H. (1980), *Ramsey Theory*, Wiley-Interscience Series in Mathematics, John Wiley and Sons, New York, Chichester, Brisbane, Toronto, $ix + 174$pp.

Greene, C., Katona, G. and Kleitman, D. J. (1976), Extensions of the Erdős, Ko, Rado theorem, *Studies in Appl. Math.* **55**, 1–8.

Greene, C. and Kleitman, D. J. (1976), The structure of Sperner k-families, *J. Combinatorial Theory (A)* **20**, 41–68.

Griggs, J. R., Stahl, J. and Trotter, W. T. (1984), A Sperner system on unrelated chains of subsets, *J. Combinatorial Theory (A)* **36**, 124–127.

Hajnal, A. and Rothschild, B. (1973), A generalization of the Erdős-Ko-Rado theorem on finite sets, *J. Combinatorial Theory (A)* **15**, 359–362.

Hall, P. (1935), On representatives of subsets, *J. London Math. Soc.* **10**, 26–30.

Hall, M. Jr. (1967), *Combinatorial Theory*, Blaisdell Publ. Co. Waltham, Mass., Toronto and London, $x + 310$pp.

Harper, L. H. (1964), Optimal assignments of numbers to vertices, *SIAM J. Appl. Math.* **12**, 131–135.

Harper, L. H. (1966), Optimal numberings and isoperimetric problems on graphs, *J. Combinatorial Theory* **1**, 385–394.

Harper, L. H. (1967), Maximal numberings and isoperimetric problems on cubes, in *Theory of Graphs*, International Symposium, Rome, 1966, Gordon and Breach, New York and Dunod, Paris, pp. 151–152.

Hart, S. (1976), A note on the edges of the n-cube, *Discr. Math.* **14**, 157–163.

Holley, R. (1974), Remarks on the FKG inequalities, *Comm. Math. Phys.* **36**, 227–231.

Horn, A. (1955), A characterization of unions of linearly independent sets, *J. London Math. Soc.* **30**, 494–496.

Kalai, G. (1984), Weakly saturated graphs are rigid, in *Convexity and Graph Theory* (M. Rosenfeld and J. Zaks, eds.), *Ann. Discr. Math.* **20**, 189–190.

Kalai, G. (1985), Hyperconnectivity of graphs, *Graphs and Combinatorics* **1**, 65–79.

Karpovsky, M. G. and Milman, V. D. (1978), Coordinate density of sets of vectors, *Discrete Math.* **24**, 177–184.

Katona G. O. H. (1964), Intersection theorems for systems of finite sets, *Acta Math. Acad. Sci. Hungar.* **15**, 329–337.

Katona, G. O. H. (1966), On a conjecture of Erdős and a stronger form of Sperner's theorem, *Studia Sci. Math. Hungar.* **1**, 59–63.

Katona, G. O. H. (1968), A theorem on finite sets, in *Theory of Graphs* (Erdős P. and Katona G. O. H., eds.), Akadémiai Kiadó, Budapest, pp. 187–207.

Katona, G. O. H. (1972), A simple proof of the Erdős-Ko-Rado theorem, *J. Combinatorial Theory (B)* **13**, 183–184.

Katona, G. O. H. (1975), The Hamming-sphere has minimum boundary, *Studia Sci. Math. Hungar.* **10**, 131–140.

Katona, G. O. H., Nemetz, T. and Simonovits M. (1964), On a problem of Turán in the theory of graphs (in Hungarian), *Mat. Lapok* **15**, 228–238.

Kirkman, T. P. (1847), On a problem in combinations, *The Cambridge and Dublin Math. J.* **2**, 191–204.

Kirkman, T. P. (1850a), Query VI, Lady's and Gentleman's Diary, 48.

Kirkman, T. P. (1850b), On triads made with fifteen things, *London, Edinburgh and Dublin Phil. Mag. and J. Science* **37**, 169–171.

Kirkman, T. P. (1850c), Note on an unanswered prize question, *Cambridge and Dublin Math. J.* **5**, 255–262.

Kleitman, D. J. (1965), On a lemma of Littlewood and Offord on the distribution of certain sums, *Math. Z.* **90**, 251–259.

Kleitman, D. J. (1966a), Families of non-disjoint subsets, *J. Combinatorial Theory* **1**, 153–155.

Kleitman, D. J. (1966b), On a combinatorial conjecture of Erdős, *J. Combinatorial Theory* **1**, 209–214.

Kleitman, D. J. (1968), Maximum number of subsets of a finite set no k of which are pairwisely disjoint, *J. Combinatorial Theory* **5**, 157–163.

Kleitman, D. J. (1970), On a lemma of Littlewood and Offord on the distributions of linear combinations of vectors, *Adv. Math.* **5**, 155–157.

König, D. (1931), Graphs and Matrices (in Hungarian), *Mat. és Fiz. Lapok* **38**, 116–119.

Kruskal, J. B. (1963), The number of simplices in a complex, in *Mathematical Optimization Techniques*, Univ. California Press, Berkeley, pp. 251–278.

Larman, D. C. and Rogers, C. A. (1972), The realization of distances within sets in Euclidean space, *Mathematika* **19**, 1–24.

Liggett, T. M. (1977), Extensions of the Erdős-Ko-Rado theorem and a statistical application, *J. Combinatorial Theory (A)* **23**, 15–21.

Lipski, W. (1978), On strings containing all subsets as substrings, *Discrete Math.* **21**, 253–259.

Littlewood, J. E. (1982), *Collected Papers of J. E. Littlewood*, 2 vols, Oxford University Press, Oxford *xxxvii* + 1675pp.

Littlewood, J. E. and Offord, A. C. (1943), On the number of real roots of a random algebraic equation (III), *Mat. Sbornik* **12**, 277–286.

Lovász, L. (1979), *Combinatorial Problems and Exercises*, North-Holland, Amsterdam, New York, Oxford 551pp.

Lubell, D. (1966), A short proof of Sperner's lemma, *J. Combinatorial Theory* **1**, 299.

Marica, J. and Schönheim, J. (1969), Differences of sets and a problem of Graham, *Canad. Math. Bull.* **12**, 635–637.

Mathias, A. R. D. (1968), On a generalization of Ramsey's theorem, *Notices of the American Math. Soc.* **15**, p. 931.

Menger, K. (1927), Zur allgemeinen Kurventheorie, *Fund. Math.* **10**, 96–115.

Meshalkin, L. D. (1963), A generalisation of Sperner's theorem on the number of subsets of a finite set (in Russian), *Teor. Probab. Ver. Primen.*

8, 219–220; English translation in *Theory of Probab. and its Applns* **8**, 204–205, published in 1964.

Nash-Williams, C. St. J. A. (1965), On well quasi-ordering transfinite sequences, *Proc. Camb. Phil. Soc.* **61**, 33–39.

Nash-Williams, C. St. J. A. (1967), An application of matroids to graph theory, in *Theory of Graphs*, International Symposium (Rome), Gordon and Breach, New York and Dunod, Paris, pp. 263–265.

Paley, R. E. A. C. (1933), On orthogonal matrices, *J. Math. Phys.* **12**, 311–320.

Rado, R. (1942), A theorem on independence relations, *Quart. J. Math. (Oxford)* **13**, 83–89.

Rado, R. (1962), A combinatorial theorem on vector spaces, *J. London Math. Soc.* **37**, 351–353.

Ramsey, F. P. (1929), On a problem of formal logic, *Proc. London Math. Soc.* (2) **30**, 264–286.

Ray-Chaudhuri, D. R. and Wilson, R. M. (1971), Solution of Kirkman's schoolgirl problem, *American Math. Soc. Proc.*, Symp. on Pure Maths **19**, 187–204.

Sauer, N. (1972), On the density of families of sets, *J. Combinatorial Theory (A)* **13**, 145–147.

Schönheim, J. (1974), On a problem of Daykin concerning intersecting families of sest, *Proc. British Combin. Conf. Aberystwyth 1973* (T. P. McDonough and V. C. Mavron, eds), London Math. Soc. Lecture Notes **13**, 139–140.

Seymour, P. D. (1973), On incomparable collections of sets, *Mathematika* **20**, 208–209.

Shearer, J. B. and Kleitman, D. J. (1979), Probabilities of independent choices being ordered, *Studies in Appl. Math.* **60**, 271–276.

Shelah, S. (1972), A combinatorial problem; Stability and order for models and theories in infinity languages, *Pacific J. Math.* **41**, 247–261.

Silver, J. (1970), Every analytic set is Ramsey, *J. Symbolic Logic* **35**, 60–64.

Sperner, E. (1928), Ein Satz über Untermengen einer endlichen Menge, *Math. Z.* **27**, 544–548.

Steiner, J. (1853), Combinatorische Aufgabe, Crelle's J. f. d. reine und angewandte Mathematik **45**, 181–182.

Tarján, T. G. (1975), Complexity of lattice-configurations, *Studia Sci. Math. Hungar.* **10**, 203–211.

Turán P. (1941), On an extremal problem in graph theory (in Hungarian), *Mat. Fiz. Lapok* **48**, 436–452.

Tuza, Zs. (1984), Helly-type hypergraphs and Sperner families, *Europ. J. Comb.* **5**, 185–187.

van der Waerden, B. L. (1927), Beweis einer Baudetschen Vermutung, *Nieuw Archief voor Wiskunde* **15**, 212–216.

Welsh, D. J. A. (1970), On matroid theorems of Edmonds and Rado, *J. London Math. Soc.* **2**, 251–256.

Welsh, D. J. A. (1976), *Matroid Theory*, LMS Monographs, No. 8, Academic Press, London, New York, San Francisco $xi + 433$pp.

Yamamoto, K. (1954), Logarithmic order of free distributive lattices, *J. Math. Soc. Japan* **6**, 343–353.

INDEX